AF412409

Current Indications for Growth Hormone Therapy
2nd, revised edition

Endocrine Development

Vol. 18

Series Editor

P.-E. Mullis Bern

Current Indications for Growth Hormone Therapy

2nd, revised edition

Volume Editor

Peter C. Hindmarsh London

5 figures and 18 tables, 2010

Basel · Freiburg · Paris · London · New York · Bangalore · Bangkok · Shanghai · Singapore · Tokyo · Sydney

Endocrine Development

Founded 1999 by Martin O. Savage, London

Prof. Peter C. Hindmarsh
Developmental Endocrinology Research Unit
Institute of Child Health
London, UK

Library of Congress Cataloging-in-Publication Data

Current indications for growth hormone therapy / volume editor, Peter C.
Hindmarsh. -- 2nd, rev. ed.
 p. ; cm. -- (Endocrine development, ISSN 1421-7082 ; v. 18)
 Includes bibliographical references and indexes.
 ISBN 978-3-8055-9194-2 (hard cover : alk. paper)
 1. Somatotropin--Therapeutic use. I. Hindmarsh, P. C. (Peter C.) II.
Series: Endocrine development, v. 18. 1421-7082 ;
 [DNLM: 1. Growth Disorders--drug therapy. 2. Hormone Replacement
Therapy--methods. 3. Human Growth Hormone--deficiency. 4. Human Growth
Hormone--therapeutic use. W1 EN3635 v.18 2010 / WK 550 C976 2010]
 RM291.2.S6C87 2010
 615'.363--dc22

                                        2010003907

Bibliographic Indices. This publication is listed in bibliographic services, including Current Contents® and PubMed/ MEDLINE.

© Copyright 2010 by S. Karger AG, P.O. Box, CH–4009 Basel (Switzerland)
www.karger.com
Printed in Switzerland on acid-free and non-aging paper (ISO 9706) by Reinhardt Druck, Basel
ISSN 1421–7082
ISBN 978–3–8055–9194–2

Contents

Preface

This is the second edition of 'Current Indications for Growth Hormone Therapy'. Since the first edition there has been a dramatic increase in our understanding of various aspects of the growth hormone system particularly from the molecular standpoint. To keep the reader up dated on this we have revised the chapter on the diagnosis of growth hormone deficiency and introduced a new chapter on the impact of molecular variants of the growth hormone receptor on the growth response to growth hormone therapy. Understanding the evidence base for the use of growth hormone remains the cornerstone of this book hence the chapter by Chris Kelnar on the evidence base and the statistical reasoning behind trial design by Vern Farewell. These chapters are important in understanding the comments made in the clinical applications chapters which consider longer term data on growth hormone usage which it has to be said has produced mixed results that warrant more detailed trials in the future.

It is 50 years since the description of the first use of growth hormone therapy in the *New England Journal of Medicine*, so it is timely for us to review where we are in that field and it was highly appropriate that Michael Ranke took the task of reviewing this area for us. Finally, we need to be mindful that our patients do grow up and need long-term follow-up under adult endocrinologists and Nelly Mauras provides us with a view on what we should be doing during this important phase of development.

Inevitably, medicine will face many challenges in the field of quality care as we go forward over the next 10 years. We know that growth hormone is effective in a number of situations so our focus will now shift towards demonstrating that such interventions actually benefit the patient and/or society. We start this discussion in this book but this is only a first attempt at a complex area that we will all need to engage in as responsible clinicians.

Peter C. Hindmarsh, London, UK

Hindmarsh PC (ed): Current Indications for Growth Hormone Therapy, ed 2, revised.
Endocr Dev. Basel, Karger, 2010, vol 18, pp 1–22

Clinical Trials: Planning and Analysis

Vernon T. Farewell[a] · Richard J. Cook[b]

[a]MRC Biostatistics Unit, Institute of Public Health, Cambridge, Cambridge, UK, and [b]Department of Statistics and Actuarial Science, University of Waterloo, Waterloo, Ont., Canada

Abstract

With specific reference to studies of growth hormone therapies, this chapter re-examines some basic principles in clinical research and considers some of the more complex aspects of trial design. The statistical structure which underpins trial design is characterized with specific attention given to samples size, subgroup analyses, multicentre trials and the role and implementation of randomization. Topics related to outcomes in studies of growth hormone therapy are discussed, including the possible value of surrogate responses, longitudinal outcomes and their analysis, and multiple outcomes. Additional design issues addressed are multiarmed trials, factorial designs and study monitoring. Brief comments are offered on the interpretation of negative trials and on non-inferiority trials.

Introduction

The randomized clinical trial is widely regarded as the optimal study design by which to investigate and communicate the utility of new treatments. Randomized trials control for confounders, facilitate causal inferences, and generate unbiased estimates of treatment effect for the relevant patient population. The classic form of a clinical trial involves a group of patients at a clinical research centre who agree to be randomly assigned to one of two treatments. After a relatively short period of follow-up, the 'outcome' of the treatment is available for enough individuals that a comparison can be made as to the relative effectiveness of the two treatments.

Many factors conspire to make this classic picture unrealistic, but the advantages of this randomized clinical trial design cannot be ignored. Indeed this simplistic design discussed above serves as a paradigm with which more complex clinical study designs should be compared. Investigations which involve growth hormone therapy present some specific challenges to the classic paradigm. The purpose of this chapter is to re-examine some basic principles of clinical research, consider some of the more complex aspects of trial design, and make specific references to how these pertain to studies of growth hormone therapies.

Statistical Structure for Planning of Trials

General Remarks
The primary aim of a clinical trial is to provide a convincing answer to an important question regarding a therapeutic intervention. A trial which cannot do this is both a waste of resources and, in our view, unethical. Therefore, these aims must be kept clearly in mind when deciding whether to initiate a trial. Whether these aims are fulfilled will depend on the choice of treatments, outcome measures, patient populations, and a variety of other concerns, including, of course, whether there is in fact an underlying effect of therapy.

Much of the planning of a clinical trial is based on the statistical significance test that is to be carried out upon completion of the trial to determine whether convincing evidence of a difference between the treatments under study has emerged. We focus here on the case of a trial comparing an experimental treatment to a standard control intervention. This will usually, although not always, be a significance test of the so-called *null* hypothesis of no treatment difference with respect to the outcome of interest. Integral to the notion of this test is the concept of a *clinically important treatment effect*. This represents the difference in the mean response, say, which would be of clinical importance to the medical community, and which the trial should have a substantial chance of detecting. If the null hypothesis is *rejected* then it is assumed that the conclusion will be that the treatments are in fact different.

If the null hypothesis is not rejected, however, then it must be concluded that there is insufficient information to claim a treatment difference with respect to the response of interest. That is, failure to reject the null hypothesis does not imply that the treatments are therapeutically equivalent. In this sense there is an asymmetry to the problem since data which leads to rejection of the null hypothesis provides evidence that the two treatments differ, but data which does not lead to rejection of the null hypothesis does not provide evidence that they are of the same therapeutic value. This point will be expanded upon in the section on noninferiority trials which appears later in this chapter.

Within this framework it is desirable to control the chance of making incorrect conclusions, which is formalized in terms of *error rates*. A type I error corresponds to rejection of the null hypothesis when the treatments do not in fact differ, and a type II error corresponds to acceptance of the null hypothesis when the treatments do differ. The type I and II error rates represent the frequency with which these errors would be committed in a hypothetical series of replicated studies. The level of the significance test is denoted by α and reflects the strength of evidence required to reject the null hypothesis; it determines the type I error rate, and is often 5%. The *power* of the study is the term used for the calculated probability that the null hypothesis will be rejected when some specified treatment difference exists, usually the clinically important treatment effect. By definition then, the power of a study is the complement of the type II error rate, often denoted β, so the power is 1 – (probability of a type II error) = 1 – β. For example, if the type II error rate is known to be 20%, then the power of the study is 80%.

The description above provides the structure within which most clinical trials are planned. In particular, sample size calculations, an important design consideration to any trial, rely heavily on this formulation. Very few trials require complex sample size calculations since consideration of a simplified version of the proposed analysis will usually be sufficient to answer one of the two questions towards which sample size calculations are directed. These are: (1) How many subjects should participate in the clinical trial? (2) Is the trial worth doing if only n (a fixed number) of patients are available? Choosing conservative estimates of the clinically important treatment effect is often advisable since it will lead to comparatively large trials. Studies of inadequate size may ultimately confuse rather than clarify.

Many of the topics which we will subsequently discuss can be addressed through what are termed Bayesian statistical methods. These require specification of a probability distribution for the treatment effect in a trial, which represents a prior state of knowledge at the initiation of the trial. Interesting work has been directed at consideration of a range of possible prior 'opinions' about the treatment effect [1]. For example, the design and monitoring criteria for a trial can be considered in the context of the need to satisfy both a physician sceptical of a treatment effect emerging and one optimistic of a treatment benefit.

There is growing interest in this intuitively appealing paradigm. While where statistical detail is necessary we adopt the widely used frequentist approach, many of the issues we will discuss are of a general practical nature. They are, therefore, relevant regardless of the framework adopted for statistical inference.

Establishing the Number of Subjects Required

Assume a continuous outcome measure is adopted in a clinical trial, such as the increase in height observed over a particular period of follow-up. Let μ_T and μ_C denote the mean increase in height in the experimental and control groups respectively and assume that the variability in the increase in height is common between groups and is denoted by σ^2. We may choose to test whether there is a difference in the mean increase in height between two treatment groups. Let $\Delta = \mu_T - \mu_C$ denote the minimal clinically important difference between the treatment groups. If the aim is to carry out a two-sided test of the null hypothesis of no difference in the mean increase in height, and we desire $100(1 - \beta)\%$ power to detect this effect, then the required sample size per group is

$$n \geq \frac{2\sigma^2(z_{\alpha/2} + z_\beta)^2}{\Delta^2}$$

In this formula, $z_{\alpha/2}$ is the percentage point of the standard normal distribution that corresponds to the significance level below which the test will be regarded as significant. For a 5% test, it is the 97.5% percentage point or 1.96. The percentage point z_β fixes the power of the test with, for example, 90% power requiring the use of the 10% percentage point or 1.64. As can be seen from this formula, studies involving more variable outcomes (larger values of σ^2) will require more subjects. Moreover, studies aiming to detect small differences between groups (i.e. studies where Δ is small) will require more subjects.

Frequently subjects withdraw from studies during the period of follow-up, which can mean that responses are unavailable for them. For example, if it is anticipated that 10% of the recruited subjects will withdraw from the study, then it would be wise to randomize $n' = n/0.9$ subjects to ensure enough subjects can provide useful responses that power objectives can be met. It must be remembered however that if withdrawals represent a particular subset of patients, perhaps those with poorer outcomes, their exclusion from analyses may bias the trial's conclusions. The aim should therefore always be to keep withdrawals to a minimum.

The above sample size formula can be easily adapted for settings with binary, count, or time to event responses. Meinert [2] gives a survey of these formulas and other issues in planning studies.

In some settings it can be difficult to obtain estimates of the parameters required for sample size calculation, such as the variance, and so pilot studies are often useful to collect preliminary data on the outcomes in the patient population of interest. These data may be used to estimate the variance which may in turn be used to plan the sample size of the full randomized trial. Pilot studies also provide information on the feasibility of implementing the intervention, provide a dry run of data collection, and give some sense of the likely rate of subject accrual in the full study which can help in setting timelines.

When subjects are difficult to recruit, the intervention is extremely costly or the required follow-up is very long, it may be undesirable to delay initiation of the main study by conducting a pilot study. In this case, so-called two-stage adaptive designs are appealing. With increasing pressure to rapidly identify effective therapies in clinical research, there has recently been an increased interest in the use of adaptive designs which facilitate sample size re-estimation based on preliminary data. Wittes and Brittain [3] were among the first to propose use of an 'internal pilot' study which is motivated by the need to eliminate the delays resulting from a separate external pilot study. Trials with internal pilot studies involve two stages. First tentative estimates of design parameters are specified for a preliminary sample size calculation. After some data have been collected on recruited patients (stage I data), estimates of parameters are obtained and these new estimates are used to carry out refined sample size calculations. The estimates based on the stage I data are known to be appropriate for the full trial since patient recruitment is identical, treatment is administered similarly, and responses are measured in the same way as in the full trial [4]. Upon completion of accrual based on the revised estimate, analyses are carried out for testing treatment effects [5] in stage II.

A considerable amount of research has recently taken place on the development of methods for two-stage designs with sample size re-estimation at the first stage [6].

Subgroup Analyses

Given the resources and time invested and the extent of data typically collected in moderate to large-scale clinical trials, it is natural to want to address a relatively large number of questions from such a study. Such a rationale often motivates subgroup analyses.

Yusuf et al. [7] distinguish between proper and improper subgroups. They define a proper subgroup as a collection of patients characterized by one or more baseline variables which reflect patient characteristics that are specified and measured before randomization and hence cannot be affected by treatment. An improper subgroup is defined as a collection of patients characterized by a variable or variables observed after treatment is initiated and hence is potentially affected by treatment. Analyses of improper subgroups are fraught with difficulties in terms of causal inferences since it is difficult to attribute the observed effect to treatment or the subgroup characteristics. As a result, we will subsequently consider only proper subgroup analyses and will omit the term 'proper'.

The first important consideration in subgroup analyses is to recognize the appropriate question to ask. In a clinical trial, a treatment is usually being given to a class of patients for which it is expected, or hoped, to be effective. At the analysis stage, the first question is whether the treatment demonstrates this effectiveness. Two secondary subgroup questions might be posed. This first is 'What is the treatment effect in particular subgroups of subjects'? The second is 'Does the effectiveness of the treatment vary across different subgroups of trial participants'? The first question is what is typically addressed by carrying out a formal analysis on a select group of patients to estimate a subgroup-specific treatment effect. This analysis strategy has numerous difficulties associated with it, in part due to the nature of the question. These subgroup analyses are typically viewed as corresponding to the analysis of a 'trial within a trial'. It is tempting to think of these findings as being most relevant for the corresponding subpopulation of patients. However, trials are not typically designed to test for treatment differences in subgroups of patients, so inadequate sample sizes lead to inadequate power and very wide confidence intervals for the estimated treatment effect. In some cases, a comparison of treatment groups for a subgroup may be specified as of interest a priori and in such cases sample size calculations may be carried out to meet power requirements for specific subgroups, but this is relatively uncommon.

In trials with an overall significant result, upon failure to reject the subgroup-specific null hypothesis, it is not always clear if one should use the overall estimate and claim a statistically significant benefit, or claim inadequate evidence that there is an effect. Moreover, there are no controls in place to limit the number of subgroup analyses one might conduct, and with a large number of tests there is an increasing risk of false positive conclusions [8].

The second question however, while related to the latter, is both more appropriate and more easily addressed. In particular, the test for subgroup differences is usually based on the test for an interaction effect in a regression model. It can be carried out whether the overall test of treatment effectiveness is significant or not. A failure to identify significant interaction terms would suggest the overall estimate is appropriate for all subgroups defined by the variables of interest. One is therefore not left with the problem of inconsistent overall and subgroup-specific estimates and conclusions. This feature is consistent with recommendations in Yusuf et al. [7] who recommend emphasizing the overall estimate of treatment effect in general.

An excessive number of subgroup comparisons should be avoided, even if multiplicity concerns are addressed. In addition, while it is entirely appropriate to undertake subgroup analyses at the end of a trial and to look for interesting findings, this must be viewed as an exploratory procedure and best serves, and should be presented, as a hypothesis generating activity for subsequent studies.

Advantages of Multicentre Trials
There are situations in which small, usually single-centre, clinical trials are unavoidable. However, in general there are hazards to small trials. Matthews and Farewell [9], for example, consider the situation when the standard response rate with the standard treatment of a disease is 30%. Assume that new drugs are being developed and that 80% of new drugs are no more effective than the standard and the remaining 20% have a response rate of 50%. Assume further that there are 20,000 potential participants for trials worldwide and that all trials are of the same size and will test for a treatment effect at the 5% level.

If trials of size 400 were undertaken, then 2 of 40 trials involving new treatments with no improvement are expected to generate a 'significant' result while virtually all 10 of the trials of improved treatments should be 'significant'. Thus the ratio of expected false positive to true positive trials will be 0.20. On the other hand, if trials of size 25 are undertaken, then 32 of 640 trials of 'no improvement' treatments and 43 of 160 trials of more effective treatments are expected to be significant. This leads to a ratio of 0.74 of false positives to true positives. It can be seen therefore that the quality of information arising from clinical trials is much greater when larger trials are the norm.

To undertake properly sized trials for a rare disorder, it is almost essential to have a multicentre trial. A single clinical centre will see only a small number of patients with the rare condition but combining patient populations from a variety of centres will be sufficient to make clinical trials feasible. To undertake a multicentre trial, it is necessary first to get a group of investigators in the different centres to agree on a question of interest and to adopt a common treatment protocol. Further, they must agree to provide patient data in a standardized manner. Generally one, or a small number, of clinical investigators will take the lead in recruiting centres for a trial and for setting trial procedures. Some sort of executive committee will be established to oversee the trial and usually a statistical centre for data management and analysis will be established.

Some aspects of a 'good' clinical trial may be more difficult to ensure in a multicentre trial, however others may be easier. For example, good quality data will usually require careful development of common data collection forms to be used in all the centres, with training for their completion undertaken in each site. The forms will be sent on a continuing basis to the statistical centre where they will be checked for completeness and consistency. This requires considerable effort. On the other hand, a multicentre trial generates, from its very nature, a generalizability not possible with a single-centre trial. The implementation of the protocol in different centres with varying patient populations does increase confidence that trial results will be relevant more generally.

Prior to a general discussion of monitoring of clinical trials later, the issue of accrual in multicentre trials deserves particular mention here. In a single-centre trial, investigators will quickly note if recruitment is consistent with their pre-trial expectations. In a multicenter trial, individual centre investigators may not be able to put their recruitment figures in context. Thus, there must be ongoing central monitoring of the recruitment pattern in a multicentre trial. A consistently small deficit in recruitment across many centres may result in a trial taking significantly longer than was envisaged. Equally, a few centres not providing patients may suggest that there are centre-specific difficulties. This may lead to problems being rectified or new centres being recruited to justify the trial's continuation.

Randomization

Clinical trials almost always represent a trade-off between individual and collective ethics. We will not discuss this in general but random treatment assignment, in particular, is a recurring topic of discussion [10–14]. There are considerable advantages to the use of random treatment assignment in the evaluation of therapeutic interventions. It has been clearly argued in the past [15] that randomization in a scientific experiment serves two primary purposes. The first purpose is to ensure that the observed treatment effect provides a 'good' (unbiased) estimate of the true treatment effect. The fact that a group of patients on one treatment is observed to do better than a group on an alternative treatment is of value only if it can then be said that the difference is attributable to the different treatments and not to something else. If a non-randomized design is used then it must be argued that the method of treatment assignment is very unlikely to be responsible for the treatment effect. But, as Cox [15] indicates, this type of statement has no measurable uncertainty. Thus randomization can be seen as the means to establish that the treatment *caused* the difference observed. The second purpose is to provide a distributional structure which facilitates certain types of statistical inference. This is more of a mathematical feature of randomization and so does not carry much weight on clinical or ethical grounds. From the medical perspective the first purpose is undoubtedly of primary importance.

It is sometimes argued that, with the advent of modern statistical methods which adjust treatment comparisons for other prognostic factors (e.g. advanced regression methodology), randomization is unnecessary. However, not only is there the potential for unknown or unspecified prognostic factors to obscure treatment effects, but regression methods are based on empirical models which, while extremely useful in general, are not expected to precisely model the medical process under investigation. Thus, anything other than random treatment assignment leaves so much opportunity for differential treatment strategies and other non-random variation from patient to patient that it is virtually impossible to attribute an observed effect solely to the different treatments with any degree of confidence. More generally, any claim that a particular means of inference mitigates the need for randomization [e.g. 11] should be of concern to medical researchers. Whatever the method of statistical inference, randomization retains its primary role in ensuring that the specific question being addressed in the analyses is in fact the question of medical interest.

An issue which arises frequently in the implementation of randomization is the stratification of treatment assignment within patient subgroups. For example, in trials of growth hormone therapy it may be felt sensible to stratify on parental height to avoid an imbalance in this factor between treatment groups. While arguments have been made that statistical methods exist to adjust for any imbalance, these are usually model based and, again, may not be convincing to a wide audience. It is sensible therefore to stratify on factors known to be of particular importance to the outcome. Reasonable balance on other factors will usually be achieved through the randomization. Excessive stratification should be avoided however as it is complicated, usually unnecessary, and may, in fact, result in poor balance if strata become too small.

While it is generally agreed that excessive stratification is to be avoided, an alternative to stratified randomization does exist. The advent of widely available computing resources allows use of a technique termed *minimization*. Minimization aims to provide an effective randomization scheme when there are more than two or three prognostic factors on which stratification might be appropriate and therefore the risk of stratification leading to poor balance exists.

Essentially, minimization does not ensure balance within each of the potentially many strata that are defined by all possible combinations of the relevant prognostic factors. Instead, it only aims to ensure that when each prognostic factor is looked at individually, there is appropriate balance between treatment assignments. The balance that has been achieved prior to a new randomization being required is examined and the probability of assignment to treatments is then specified for the new randomization in a way that is likely to reduce any imbalance present. The algorithm to specify the randomization probabilities is relatively complex and thus requires access to computer resources. It is this complexity which provides protection against selection bias.

From the design perspective, a moderate level of stratification may be sufficient and an easier option for many trials. Also, there is some debate concerning methods of analysis when minimization is used but often these are not of practical importance. Nevertheless, the use of either stratification or minimization is likely acceptable but note that the analysis of a trial must take account of factors used in the randomization procedure.

For factors not balanced at randomization, it remains common, in reporting the final results of a trial, to present a table comparing the distribution of factors across randomized groups. While this may be reassuring when it demonstrates balance in risk factors across groups, the use of statistical tests to examine any differences should usually be avoided at this stage. The logic of a statistical test is that a small p value may result from either a difference in the compared groups or chance. When a comparison is 'significant', it is usually taken to suggest a difference. In the comparison of randomized groups, however, any significance must be due to chance. If the imbalance is not simply a rare event, then the most likely reason is a compromised randomization process. This possibility must be considered as it may cast doubt on the trial management more generally. However, this should not be discovered at the time of a final analysis and this is the reason that the presentation of significance tests is problematic. If an imbalance has occurred by chance, then there can be debate as to whether this requires adjustment in the final analyses. The pragmatic approach will be to perform adjusted and unadjusted analyses and only if qualitative differences emerge will interpretation of the trial be made more difficult.

Outcomes in Studies of Growth Hormone Therapy

Overview
Studies involving growth hormone therapy present a number of challenges regarding outcome selection. The reason relates, in part, to the fact that the outcome of usual interest is final height, which is difficult to observe in the relatively short time horizon of a typical prospective study. Instead, measures such as height velocity, pubertal stages, and bone age are utilized as outcomes. Frequently outcomes are adopted which are designed to be predictive of final height such as height for bone age score [16]. See Knudtzon et al. [17] for a variety of other measures. Furthermore, some assessment must be made of the safety of treatments under study. Adverse outcomes which are often examined in studies of growth hormone therapy include glucose homoeostasis, excess bone growth, hypertension [18], and painful injections [19]. From these examples, it is clear that the notion of a response to treatment is multifaceted. Indeed, in many cases it is difficult to identify a single variable that captures all relevant aspects of the response to treatment. We therefore give some guidelines on the selection of primary response variables and make related comments in the following section.

Selection Criteria for Primary Outcomes
Meinert [2] suggests the following five criteria for primary outcomes in clinical trials. They must be (i) clinically relevant, (ii) easy to diagnose or observe, (iii) chosen before the start of data collection, (iv) capable of being observed independent of the treatment assignment, and (v) free of measurement or ascertainment error. We comment on the first criterion in the following section. The second criterion is given for practical reasons and the third to ensure outcomes are not selected after evidence of apparent treatment effects has emerged. This is to ensure validity, i.e. that clinical inferences are not based on possibly spurious effects. The fourth criterion is specified to ensure that there is no bias induced in the examination of the outcomes by investigators. Blinded outcome assessment is widely known to be an important feature of a clinical trial. Chatelain et al. [19, p. 1455], for example, state that in their study bone age was assessed in a blinded fashion.

Regarding the fifth criterion, we note that responses in growth hormone therapy are frequently based on measurement systems which have measurement error associated with their use. The assumption that a reliable method of determination is available is key to outcome assessments in growth hormone therapy trials as they are often based on imperfect measurement systems. Furthermore, these measurement systems are often used repeatedly over time to calculate height velocities, so measurement error can have a compounding effect.

To minimize the impact of measurement error, and hence to increase precision of the resulting estimates of treatment effect and the power of the study, Chatelain et al. [19] used the average of three repeated measurements of height for each patient at a fixed point in time and Rose et al. [20] achieved still greater precision by using the average of ten measurements. Failure to address outcome measurement error in the design and analysis of a clinical trial, will result in reduced power, and attenuated (biased) measures of treatment effects.

In general, when a measurement system is being considered, the level of reliability demonstrated in published studies should be examined. There are no satisfactory benchmarks by which to judge reliability in a standard fashion, making such assessments somewhat arbitrary. This does not, however, undermine the importance of these types of assessments, particularly when multiple outcomes are of interest. The representativeness of the physician and patient population participating in the reliability studies should be considered. Often physicians taking part in reliability studies are specially trained and are not representative of the general population of relevant physicians. Similarly, patients may not constitute a random sample and so may not be representative either. These points typically receive very little attention and yet have the potential for significant impact in the evaluation of treatments.

The existence of a single primary outcome that adequately characterizes response to treatment considerably simplifies the problem of designing a clinical trial. While there is an arbitrariness to the clinical judgment used in choosing a single outcome, in some cases it offers a pragmatic solution. Specifically, the primary response is central in the selection of the trial design and sample size.

One approach to identifying suitable primary response variables, which has much to recommend it when feasible, is to recognize that there is likely to be a high correlation between a given collection of outcomes. As a result, with little loss in information, one can often choose a single outcome variable on which most individuals would expect any effective treatment to induce a difference. Secondary outcome variables should be measured but the main analysis and clinical inferences of the trial should be based on the primary response. Secondary analyses should be conducted with a view of providing confirmatory and supportive evidence of the overall conclusions. The drawback to this solution is that unless there is general agreement on the choice and definition of a primary response, the interpretation of study results could vary in the medical community and the ability to compare results across trials might be compromised.

Finally, it is worth noting that as well as specifying the outcome for a clinical trial, the method of analysis should also be specified before the trial begins. Usually the prime consideration in this is the type of outcome variable and the extent to which factors other than treatment must be considered in the analysis. The choice of analysis method also suggests the appropriate type of sample size calculation for planning the trial.

The first criterion for primary outcomes specified by Meinert [2] is particularly relevant in studies of growth hormone therapy for children with intrauterine growth retardation. Growth velocity is perhaps the most commonly used outcome, but it is really only useful if it is directly related to final height attained. Chatelain et al. [19] report a waning effect of growth hormone therapy on height velocity during the second year of a 2-year treatment. Rosenfeld et al. [21] noted that by the third year of treatment with growth hormone, the mean height velocity is statistically unchanged from that at baseline prior to the commencement of treatment [18]. This raises questions about the relevance of height velocity as a meaningful response in growth hormone therapy trials.

To illustrate the need for clinical relevance and prognostic value, consider the developments in the evaluation of early HIV/AIDS therapy trials. In many early AIDS trials of antiretroviral drugs, changes in CD4 counts were significantly better for individuals on active treatment than controls. As CD4 cell counts reflect immune function, they were argued to be clinically relevant. However, it was only if these effects translated into better survival or quality of life that there would be a broad agreement on the treatment's value. It emerged that, in terms of prognostic value, CD4 cell counts are very strongly predictive for the development of AIDS and survival in HIV-infected individuals, but they do not reflect enough of the variation to be 'surrogates' for these longer term outcomes. Thus considerable caution should be exercised if variables are being chosen as outcomes because of early availability, convenience in measurement, or for their statistical properties.

Longitudinal Measurements

In order to ensure meaningful treatment comparisons, it is essential for studies of growth hormone therapies to involve prolonged observation of patients. When study designs involve regular follow-up visits in order to monitor various developmental aspects, a considerable amount of data may be collected. It is natural to then want to fully exploit this data in order to answer questions of clinical importance

To be precise, here we introduce some simple notation. Let y_t represent the height of subject at time t, $t = 0, 1, 2,..., T$, where $t = 0$ represents the time of randomization, and $t = T$ represents the time at study completion. Frequently, there will be associated with each follow-up visit, a set of additional subject characteristics, which we represent by x_t, $t = 0, 1, 2,..., T$, but here we will focus on simple treatment comparisons for simplicity, which will be assumed to remain fixed over time.

Perhaps the simplest and most common approach to analysing longitudinal data in studies of growth hormone therapy is to calculate height velocity which is measured by, say, $v_T = (y_T - y_0)/T$. This forms a derived response which may then be analysed in a standard fashion based on easily interpretable treatment effects. Subject to caveats pointed out in the previous section, the treatment which is associated with, on average, an increased height velocity may be expected to be advantageous. A slightly more sophisticated approach which often warrants consideration is to form a simple linear regression model in which height velocity is the response, but the height at randomization and the treatment received serve as explanatory variables. Such a regression model takes the form

$$v_T = a + by_0 + cx + \varepsilon$$

where $x = 1$ for subjects on the experimental treatment, and is zero otherwise, and leads to what is called analysis of covariance. The term ε is used to represent random variation about the expected height velocity. The advantage of introducing the height at randomization into a regression model is that it may serve to explain variation in the growth velocities separate from that due to the treatment effect alone (i.e. subjects which are taller at randomization may tend to have slower height velocities).

An additional generalization of this model is to allow the relationship between growth velocity and height at randomization to depend on treatment. This is done by introducing an interaction term into the model. This results in the model

$$v_T = a + by_0 + cx + dxy_0 + \varepsilon$$

Minimally, a test for $d = 0$ should be undertaken before the simpler version of the model is used for adjustment purposes. However, in some cases, there may be particular interest in whether one aspect of the treatment effect is to alter the relationship between growth velocity and initial height.

The above model relies on height measurements at two time points only. This model may be the most practical for many randomized clinical trials where simple measures of treatment effect are often of interest. However, more complex models, including so-called growth curve models, are attractive if more detailed comparisons are desired regarding the growth rates over time. These models typically require a high frequency of measurements over time (i.e. a large number of follow-up visits) but may offer more insight into the nature of any treatment effect. The price for adopting such models, however, is the introduction of modelling assumptions. When these modelling assumptions appear realistic and/or can be checked, the more sophisticated models can lead to greater understanding of treatment effects.

A particular form of models which might be used for such longitudinal analysis over time is random effects models. If, for example, height is measured at various points in time, then a possible model could be written as

$$Y_{ik} = \beta_{0i} + \beta_{1i}t_k + \varepsilon_{it}$$

where Y_{ik} is the measured height for individual i at time point (often a specific age) t_k and ε_{it} represents the random variation in this observation. The model assumes that expected height for individual i can be represented as a linear function of time where the form of the linear relationship is determined by a constant term β_{0i} and a slope term β_{1i}. This form, determined by β_{0i} and β_{1i} is allowed to vary arbitrarily among individuals and it is assumed that this variation can account for the correlation between observations from the same subject.

Statistical problems can arise when estimating a large number of β parameters in some settings. An additional assumption is therefore often made that the β values arise from some common statistical distributions. Commonly, it would be assumed that the β_{0i} values arise from a normal distribution with mean a and variance σ^2_a. The β_{1i} values are assumed to derive from a normal distribution with mean $b_0 + b_1 x$, where $x = 1$ if the individual is receiving experimental therapy and $x = 0$ otherwise, and with variance σ^2_b. It is common, as well, to assume that there may be a correlation between the β_{0i} and β_{1i} values in a subject. The assumption that these values come from common distributions is why such a model is termed a 'random effects' model.

Of particular interest here is the parameter b_1 which represents how much the rate of height increase is changed with the experimental therapy. While the background rate is allowed to vary across individuals, this effect of therapy is assumed to be the same for all individuals. However, note that b_1 is now a measure of how the slope of any particular individual, with a specific value of b_{1i}, would be changed if the individual was on the experimental therapy compared with what it would be if they were not.

When interest lies in examining treatment effects on the rates at which various aspects of body development occur, another class of models, termed multistage models, are attractive. Such models are based on the assumption of an underlying progressive condition, which is quite appealing in the context of growth hormone therapy trials where subjects may progress through stages of height, bone age, or pubertal development. With such an analysis in mind, it is necessary at the design stage to specify the various stages of development if they are not already defined. Upon long-term follow-up and relatively frequent clinical examination, patients will then be observed to make transitions between the different stages.

A treatment which is associated with rapid progression through these stages may be considered beneficial in this regard. Multistage models can be used to provide estimates of the rate of transitions between the different stages.

Multiple Outcomes

It is often difficult to get general agreement on a single primary outcome variable when response to treatment is multifaceted. Even for a given outcome, there are often alternative ways of measuring it (e.g. height velocity may be measured in cm/year or as a SD score). Outcomes should generally be chosen so that each one represents a different aspect of the condition of interest, and, in totality, provide a comprehensive assessment.

The issue of multiple comparisons arises when more than one outcome variable is present. As mentioned earlier, if testing is carried out at a particular significance level, α, and if the null hypothesis is true, then the probability of a false positive result is α. If a number of significance tests are performed, all with a false positive error rate of α, then the percentage of trials involving *one or more* false positive errors will be much more than α. There are a variety of statistical methods to control the overall error rate, which is defined as the rate of making one or more false positive errors. The simplest and most common is the so-called Bonferroni adjustment which divides each significance level by the number of tests performed. This is conservative, but not grossly so, if outcomes are only moderately correlated as would often be expected.

Cox [22], however, has pointed out that probabilities pertaining to the simultaneous correctness of many statements may not always be directly relevant, particularly when a specific response is of interest. This suggests that in designing a study that formally involves two or more responses, multiplicity adjustments may not be necessary if the results of individual hypothesis tests are interpreted separately and have consequences that involve different aspects of the treatment prescription.

Another approach to multiple outcome data is to adopt a single summary response, sometimes called a 'pooled index', that combines an individual's responses. The interpretation of a treatment effect based on such indices can be difficult however. This is particularly true of indices derived from statistical models, but also for those clinically motivated. It is also possible to construct global summary statistics which are usually a weighted average of the separate outcome-specific statistics. These do not involve a summary response at the patient level so comparison of individual patients is not possible and again, interpretation is difficult as the separate outcomes often remain of interest in their own right.

Finally, we remark that discussion of significance tests is useful to highlight the logical issues. However, results of analyses should be summarized by reporting point estimates, confidence intervals, and p values. Simply reporting results as significant or non-significant is inadequate. This is particularly relevant for multiple outcomes as study designs may be inadequate to detect meaningful treatment effects on all outcomes.

Issues Related to Trial Design

The choice of treatments to be compared is a primary determinant of whether a clinical trial can provide an answer to an interesting question. It deserves careful thought and some primarily technical issues are discussed here to aid in this process. It is worth noting first, however, that whatever treatments are chosen must be sufficiently acceptable to the clinicians who will enrol patients in the trial.

Multiarmed Trials
The classic paradigm of a clinical trial involves the comparison of two treatment groups. The simplicity of the two-treatment, randomized trial is attractive. For example, a decision to compare two very different treatments will likely produce results more quickly than will the comparison of similar treatments or a larger number of treatments. However, the complexity of modern therapies means that, increasingly, clinical trials with multiple treatment arms are being designed. The different treatment arms might involve various dosages of an experimental treatment, cumulative combination therapies, or simply multiple different treatments. Designs involving multiple treatment arms may provide an efficient method of simultaneously assessing the efficacies of a variety of treatment options. For example, several active treatments can be compared to one another, and/or to a common control group receiving standard medical care [e.g. 19]. Here again, however, the issue of multiple comparisons, or multiple significance tests, arises and strategies must be adopted to address related concerns.

One approach often used for the analysis of multiarmed trials is to predicate all specific comparative analyses of treatment differences on the outcome of an 'omnibus' test for the absence of any difference between any treatment groups. These tests are very useful in many contexts, but do tend to be rather conservative if relatively few differences are both possible and of interest. An alternative strategy is to specify the comparisons of interest at the design stage of the trial. This ensures that no treatment comparison is chosen because of observed differences between treatment groups that actually may be spurious. In this case, each comparison is examined separately.

A pragmatic approach to the resulting multiplicity problem is to limit the number of comparisons considered to one less than the number of treatment arms. If the number of treatment comparisons exceeds $K - 1$ when the total number

Table 1. Factorial design for treatment of early rheumatoid arthritis involving ciclosporin and/or steroids among patients receiving methotrexate (MTX)

	MTX	MTX + ciclosporin
MTX	Regimen *A*	Regimen *B*
MTX + steroids	Regimen *C*	Regimen *D*

of treatment arms is K, then it is no longer clear that each comparison is providing separate information. In this case, the view that each significance test can be interpreted separately to assess the treatment information contained in the trial is questionable.

The challenge is then to specify the appropriate $K - 1$ comparisons of interest. Clearly, situations do arise in medical research when there is understandable vagueness about the precise comparison or comparisons that might be of greatest interest. However, such non-specific hypotheses do not reflect the situation in most phase III clinical trials.

Factorial Designs

A special case of multiple treatment arms arises when there are two or more questions of interest. For example, a clinical trial of treatment for early rheumatoid arthritis examined the effectiveness of two possible treatments, ciclosporin and steroids, either or both of which could be added to a standard treatment with methotrexate [23]. The primary questions related to the efficacy of ciclosporin and the efficacy of steroids for this condition. The schematic layout for the design is shown in table 1. In such a design, patients are randomized among four regimens, and the sample size of the study need not be much larger than that required to answer either question separately. The ciclosporin question is addressed by comparing regimen *A* to regimen *B* and regimen *C* to regimen *D*. The steroids comparison is based on *A* vs. *C* and *B* vs. *D*.

With this design, it is also possible to detect a synergistic (or antagonistic) interaction between the two additional therapies as some patients receive both. However, the power for detecting such effects will be much less than that for the main questions. Thus if such an effect is of particular interest, it will be necessary to increase the sample size.

Study Monitoring

It has long been recognized that clinical trials must be monitored on an ongoing basis. There is the need to ensure that the trial is being implemented as was expected. This includes aspects such as accrual, protocol violations, and unexpected side effects of treatment. There is sometimes the need, as well, to monitor the accumulating evidence concerning possible treatment effects as the trial proceeds. This latter task introduces some particular difficulties.

Methods for monitoring study results are typically based on carrying out repeated tests of significance on the clinical trial data as it accumulates over the course of the study. As with many of the other analysis strategies involving multiple analyses, one is faced with a resultant inflation of the type I error (or false positive) rate.

This inflation of the type I error rate from repeated significance testing on accumulating data was first clearly demonstrated by Armitage et al. [24]. Armitage [25] provided an initial solution to this problem that allows interim analyses. This approach was subsequently generalized by Pocock [26] to a group sequential procedure where the maximum number of interim analyses is fixed. A number of alternative group sequential procedures have since been developed. One that we think could be quite useful in trials of growth hormone therapies is that of Fleming et al. [27].

These procedures, usually referred to as Fleming-O'Brien procedures, specify that early analyses are undertaken with only very extreme (small) significance levels leading to consideration of stopping the trial. This has the additional advantage that, when interim analyses are carried out at extreme testing levels, then at the final analysis the significance level to be used will be quite close to the overall type I error level (usually 5%), and analyses can proceed virtually as if no early testing was done. This type of statistical monitoring is particularly well suited to trials in growth hormone therapy because, with the conditions under investigation, early stopping of the trial would only be considered in the situation of dramatic treatment differences on an outcome of interest. That is, suggestive small differences would usually motivate continuing the study to confirm the magnitude of the difference.

The decision to stop a trial is not a solely statistical one and new external information may have considerable impact, particularly in studies of a long duration. Publication of the results from concurrent similar studies, the advent of new therapies, and many other issues play a role, and this reality must be recognized. However, care must be taken to avoid unnecessary abandonment of trials which, if continued, may ultimately provide definitive information regarding treatments. For example, the continuation of a study with an accepted clinically relevant outcome should not often be affected by preliminary results from other investigations involving a surrogate response.

Often when treatment results are being monitored, and particularly if the trial is double blind, an external Data and Safety Monitoring Committee will be used to provide a recommendation to the study investigators. Such committees include ethicists, clinicians, statisticians, lawyers, community representatives and others who are not directly involved in the study and can provide a broad perspective on the trial.

Interpretation of Negative Trials

Post-Trial Power Calculations
In their book on clinical epidemiology, Fletcher et al. [28] refer to p_β, the probability of a false negative trial, and state that until recently 'it has been unusual to find a careful consideration of p_β when the results of clinical trials were presented'. They go on to say that the probability of a type II error (failing to detect a real effect) 'is the main question that should be asked when the results of a study indicate "no difference" '. Such statements are becoming common, in part because of requirements of regulatory agencies, but also because of the desire to understand the results of a trial, in particular, of the so-called negative trial with no demonstrated treatment differences. Post-hoc power calculations are not the optimal way to achieve this purpose however. We suggest that there is no question asked at the end of a trial, and for which a power calculation could provide information, that is not better addressed by the more direct means of estimation. Confidence intervals specify which values of the treatment effect are plausible given the data collected. If a value is in the confidence interval, then it cannot be ruled out. If it is not, then it is unlikely to be the true treatment effect.

Further, there is a decision-based paradigm to power calculations. The trial is envisaged as either accepting or rejecting a null hypothesis of no treatment differences. Although useful at a design stage, this paradigm is often not consistent with the scientific aims of a clinical trial. A trial should provide as much information as possible about the treatments under investigation but may not lead to an immediate, definitive, and generally accepted, treatment strategy. Note too that the analysis used to produce the confidence intervals is often far more sophisticated or detailed (for example, in reflecting patient variability and trial structure) than that typically employed in sample size and power calculations at the design stage. Cox [15] says it succinctly, 'It (i.e. *power*) is quite irrelevant in the actual analysis of data'.

Non-Inferiority Trials

The standard of care for growth hormone disorders has improved considerably over the past several decades as a result of carefully designed and executed randomized clinical trials. In some settings now the focus is on the development of new treatments which have advantages over the standard of care in terms of costs, adverse event profile, or ease of delivery [29, 30].

A new treatment of this sort must be formally compared to the standard of care with the aim of demonstrating that it is not therapeutically inferior [30–32], which means that on balance the cost, adverse event, ease of delivery and effectiveness profiles make the two treatments similar enough that one might reasonably use either one. This has led to the increased interest in non-inferiority designs [33, 34]. If non-inferiority of the experimental therapy to the active control can be claimed for a primary outcome, it may be considered a preferable treatment based on cost, convenience, or adverse event profile.

In order to design non-inferiority studies, considerable care must be taken to ensure a suitable criterion is specified for a non-inferiority claim [35]. Criteria are typically based on estimates of the effect of the active control relative to a placebo obtained from historical placebo-controlled trials of the active control. A common assumption is that this effect is stable over time and hence that the benefit over placebo that was observed in historical trials might reasonably be assumed to be present in the current planned study. Relaxation of this assumption is of course possible, and it should be considered how this might impact design.

Having specified the effect of the active control against the placebo, one must then specify the 'percentage' of this effect which must be retained by the experimental treatment in order for it to be considered non-inferior to the active control [36]. Some loss in efficacy may be tolerated if there are substantial advantages anticipated in terms of the incidence of adverse effects, costs, etc, but typically the acceptable loss is modest since the intention is to convince the broader scientific community that non-inferiority is satisfied.

Stringent methods must be used in non-inferiority trials because of the concern of 'bio-creep'. Bio-creep is the phenomenon that may arise with a series of successful non-inferiority trials. In this case, there is the possibility that slightly inferior treatments are judged non-inferior and subsequently used as active controls, leading over time to treatments being adopted that are significantly inferior to the active control treatment used in the first trial.

A variety of methods have been described for designing and analysing data from non-inferiority trials, but all methods require a reliable and valid estimate of the effect of the standard care versus a placebo treatment, the 'two confidence interval' approach [37]. For all of these approaches, notions of type I error, power, and anticipated effects arise and, as in traditional superiority studies, must be specified in order to derive the required sample size (see Wang et al. [34] or Tsong et al. [38] for recent reviews).

To date, much of the work on non-inferiority trials has focused on continuous, binary, or survival time responses. In the context of growth hormone therapy trials, tests based on features of growth curves would be natural outcomes.

Conclusion

In this chapter, we have provided an overview of a variety of issues involved in trial design, emphasized the critical role of outcome selection, and highlighted important points for consideration at the analysis stage. This cannot be seen as an exhaustive discussion of methodologic issues pertaining to studies involving growth hormone therapy. Any particular trial will have unique features making such exhaustive coverage impractical. Our objective was simply to provide some general motivation for the use of well-designed studies and for the appropriate analysis of the data obtained therein. Discussion with a statistician is recommended when a study is being designed. More generally, we hope this overview also provides a basis for further discussion of study design and analysis in this important research area.

References

1 Spiegelhalter DJ, Freedman LS, Parmar MKB: Bayesian approaches to randomized trials. J R Stat Soc A 1994;157:357–416.
2 Meinert CL: Clinical Trials: Design, Conduct and Analysis. New York, Oxford University Press, 1986.
3 Wittes J, Brittain E: The role of internal pilot studies in increasing the efficiency of clinical trials. Stat Med 1990;9:65–72.
4 Birkett M, Day SJ: Internal pilot studies for estimating sample size. Stat Med 1994;13:2455–2463.
5 Betensky RA, Tierney C: An examination of methods for sample size recalculation during an experiment. Stat Med 1997;16:2587–2598.
6 Gould AL: Sample size re-estimation: recent developments and practical considerations. Stat Med 2001;20:2625–2643.
7 Yusuf S, Wittes J, Probstfield J, Tyroler HA: Analysis and interpretation of treatment effects in subgroups of patients in randomized clinical trials. JAMA 1991;266:93– 98.
8 Cook RJ, Farewell VT: On multiplicity considerations in the design and analysis of clinical trials. J R Stat Soc A 1996;159:93–110.
9 Matthews DE, Farewell VT: Using and Understanding Medical Statistics, ed 4. Basel, Karger, 2007.
10 Barnard GA: Causation; in Kotz S, Johnson NL (eds): Encyclopedia of Statistical Sciences. New York, Wiley, 1986, vol 1, pp 387–389.
11 Royall RM: Ethics and statistics in randomized clinical trials. Stat Sci 1991;6:52–88.
12 Sprott DA, Farewell VT: Randomization in experimental science. Stat Papers 1993;34:89–94.
13 Pocock SJ, Elbourne DR: Randomized trials or observational tribulations. N Engl J Med 2000;342: 1907–1909.
14 Deeks JJ, Dinnes J, D'Amico R, Sowden AJ, Sakarovitch C, Song F, Petticrew M, Altman DG: Evaluating non-randomised intervention studies. Health Technol Assess 2003;7:iii–x, 1–173.
15 Cox DR: Planning of Experiments. New York, Wiley, 1952.
16 Stanhope R, Preece MA, Hamill G: Does growth hormone treatment improve final height attainment of children with intrauterine growth retardation. Arch Dis Child 1991;66:1180–1183.
17 Knudtzon J, Aarskog D, and the Norwegian Turner Study Group: Results of two years of growth hormone treatment followed by combined growth hormone and oestradiol in turner syndrome. Horm Res 1993;39(suppl 2):7–17.
18 Chipman JJ: Study design for final height determination in turner syndrome: Pros and cons. Horm Res 1993;39(suppl 2):18–22.
19 Chatelain P, Job JC, Blanchard J, Ducret JP, Olivier M, Sagnard L, Vanderschueren-Lodeweyckx M: Dose-dependent catch-up growth after 2 years of growth hormone treatment in intrauterine growth-retarded children. J Clin Endocrinol Metab 1994; 78:1454–1460.

20 Rose RR, Leong GM, Yanovski JA, Blum D, Heavner G, Barnes KM, Chipman JJ, Dichek HL, Jacobsen J, Oerter Klein KE, Cutler GB Jr: Thyroid function in non-growth hormone-deficient short children during a placebo-controlled double blind trial of recombinant growth hormone therapy. J Clin Endocrinol Metab 1995;80:320–324.

21 Rosenfeld RG, Hintz RL, Johanson AL, Sherman B, Brasel JA, Burstein S, Chernavsek S, Compton P, Frane J, Gotlin RW, et al: Three-year results of a randomized prospective trial of methionyl human growth hormone and oxandraolone in Turner syndrome. J Pediatr 1988;113:391–400.

22 Cox DR: A remark on multiple comparison methods. Technometrics 1965;7:223–224.

23 Choy EHS, Smith CM, Farewell VT, Walker D, Hassel A, Chau L, Scott DL: Factorial randomised controlled trial of steroids and combination disease modifying drugs in early rheumatoid arthritis. Ann Rheum Dis 2008;67:656–663.

24 Armitage P, MacPherson CK, Rowe BC: Repeated significance tests on accumulating data. J R Stat Soc A 1969;132:235–244.

25 Armitage P: Sequential Medical Trials. Oxford, Blackwell, 1975.

26 Pocock SJ: Group sequential methods in the design and analysis of clinical trials. Biometrika 1977;64:191–199.

27 Fleming TR, Harrington DP, O'Brien PC: Designs for group sequential tests Control Clin Trials 1984;5:348–361.

28 Fletcher RH, Fletcher SW, Wagner EH: Clinical Epidemiology. Baltimore, Williams & Wilkins, 1988.

29 Fleming TR: Evaluation of active control trials in aids. J Acquir Immune Defic Syndr 1990;3(suppl): S82–S87

30 Hung HMJ, Wang J, Tsong Y, Lawrence L, O'Neil RT: Some fundamental issues with non-inferiority testing in active controlled trials. Stat Med pages 2003;22:213–225.

31 Blackwelder WC: Proving the null hypothesis. Control Clin Trials 1982;3:345–353.

32 Rothman KL, Michels KB: The continued unethical use of placebo controls. N Engl J Med 1994; 331:393–398

33 Jones B, Jarvis P, Lewis JA, Ebbutt AF: Trials to assess equivalence: the importance of rigorous methods. Br Med J 1996;313:36–39.

34 Wang SJ, Hung HMG, Tsong Y: Utility and pitfalls of some statistical methods in active controlled clinical trials. Control Clin Trials 2002;22:15–28.

35 Temple R, Ellenberg SS: Placebo-controlled trials and active-controlled trials in the evaluation of new treatments. 1. Ethical and scientific issues. Ann Intern Med 2000;133:455–463.

36 Holmgren EB: Establishing equivalence by showing that a pre-specified percentage of the effect of the active control over placebo is maintained. J Biopharm Stat 1999;9:651–659.

37 Rothmann M, Li N, Chen G, Chi GYH, Temple R, Tsou H-H: Design and analysis of non-inferiority mortality trials in oncology. Stat Med 2003; 22:239–264.

38 Tsong Y, Wang SJ, Hung HMJ, Cui L: Statistical issues on objective, design and analysis of non-inferiority active-controlled clinical trial. J Biopharm Stat 2003;13:29–41.

Dr. Vernon T. Farewell
MRC Biostatistics Unit, Institute of Public Health, University Forvie Site
Robinson Way, Cambridge CB2 0SR (UK)
Tel. +441223 330371, Fax +441223 330388
E-mail vern.farewell@mrc-bsu.cam.ac.uk

Hindmarsh PC (ed): Current Indications for Growth Hormone Therapy, ed 2, revised.
Endocr Dev. Basel, Karger, 2010, vol 18, pp 23–39

The Evidence Base for Growth Hormone Effectiveness in Children

Christopher J.H. Kelnar

Paediatric Endocrinology, Section of Child Life and Health, Division of Reproductive and Developmental Sciences, University of Edinburgh, Edinburgh, UK

Abstract

Clinical medicine is a holistic attempt to provide the best care for patients. A clinician's knowledge may be biased (belief vs. knowledge), families' and patients' expectations may be unrealistic, and a 'worthwhile' outcome may be difficult to define. New evidence, which may or may not be of high quality and may or may not be rigorously evaluated by the clinician's own critical review, is then added to a set of prior beliefs to influence prescribing practice. Applying evidence to inform high-quality patient care is not straightforward. Methodological difficulties with many studies of childhood growth include underpowered studies, predicted or projected height comparisons, historical control groups, selection bias and variable treatment protocols. Many growth studies are reported based on 'predicted adult height' or other surrogate markers – even final height is not a validated proxy for psychological contentment or 'quality of life' (QoL). Because you can measure something (or think you can) does not in itself make it important and clinically relevant. Sometimes what is most important is most difficult to measure, and even when it can be, statistical significance is not the same as a clinically relevant difference. Making someone taller, even if achievable, is not an end in itself. What outcomes do we (and patients) consider clinically important and relevant? In growth hormone (GH) therapy studies should it be final height, height at or through puberty, psychological benefit (short vs. long term) or QoL improvement? How do these relate to safety issues and risks of harm? The limitations of growth screening, GH testing and our lack of information on, and understanding of, clinically relevant outcomes in response to GH therapy are discussed in these contexts.

'The questions are the same but the answers are different.'
[Allegedly said by Albert Einstein in response to a complaint from a student about why the examination questions were the same as those set the previous year]

Evidence Is Not Truth – How Do We Know What We Know?

What is the relevance of evidence-based child health (EBCH) to clinical practice and what are its limitations? [1]. One definition of EBCH is 'the integration of clinical information obtained from a patient [to which one might add 'and their carer(s)'] with the best evidence available from clinical research and experience and the application of this knowledge to the prevention, diagnosis or management of disease in that child' [2].

A clinician will have a prior view as to the likelihood of a drug providing benefit ('improved clinical outcome') for a patient. Although this will be informed by knowledge of the available scientific literature, the patient's and the family's views, etc., this already begs a number of questions: the clinician's knowledge may be biased (belief vs. knowledge); families' and patients' expectations may be unrealistic. What is a worthwhile outcome? How do you define and measure benefit(s)? New evidence, which may or may not be of high quality and may or may not be rigorously evaluated by the clinician's own critical review, is then added to this set of prior beliefs to influence prescribing practice.

Treatment comparisons based on impressions, or restricted analyses, only provide reliable information in rare circumstances when treatment effects are dramatic (e.g. hygiene for preventing tetanus in newborn babies (reported by Schleisner in 1849) and insulin in treating diabetes mellitus [3]. Virtually all therapies we use or will develop for use in the future in clinical medicine have much less dramatic effects. Unless care is taken to avoid biased comparisons, dangerously mistaken conclusions about the effects of a treatment may result.

Amongst the important attributes of a clinician is the ability to synthesize complex information – the exercise of clinical judgment against a background of imperfect knowledge, risk and uncertainty. To make a balanced judgment as to whether to prescribe a drug for a particular indication in a particular child, information is needed about both efficacy and safety and, because of the nature of most clinical trials, efficacy data are much more quickly and readily available.

There is a belief that randomized controlled trials (RCTs) are 'better' than case control or cohort studies and this is the case in looking at the 'efficacy' of a drug (see Chapter 2 for full discussion of this topic). However, the situation is different regarding harms. The possibility of unwanted outcomes (side effects) should and generally does influence decision-making about prescribing in the real world (*primum non nocere*) but there is an additional 'bias' in the nature of 'outcomes' available in the short term in favour of efficacy rather than side effects. Even large RCTs are unlikely to uncover what can be merely uncommon significant adverse side effects [4]. Put simplistically, if an event occurs only 1 in 500 patients then at least 1,500 patients will be needed to receive the treatment to be sure the effect is not occurring and even then a much larger database will be required for certainty.

In 1978, 20 years after pituitary GH was introduced to treat children with GH deficiency, Tanner [5] was still able to state that 'very occasionally antibodies develop to the injected (growth) hormone and cancel out the effects of treatment; otherwise side effects are practically unknown'. Yet within 7 years there were reports linking pituitary GH therapy with Creutzfeldt-Jacob disease [6–9].

In this context, databases such as the Kabi International Growth Study (KIGS), founded in 1987 in order to document the *safety* and efficacy (my italics) of the then new recombinant human GH, and now comprising an extremely large pharmaco-epidemiological survey (>60,000 patients worldwide), find their main justification. Even in rare conditions, the patient-years of treatment experience documented can be very considerable [10].

Databases (like RCTs) have strong and weaker points. The quality of diagnostic and monitoring data submitted can be variable and difficult to verify, and it is likely that, for this and other reasons, databases such as KIGS are particularly valuable for assessing safety issues and the prevalence/incidence of side effects, etc., rather than efficacy outcomes per se.

A balanced view of the assessment of studies reporting the effects of an intervention would suggest that observational studies contribute complementary evidence about the effects of treatment on significant adverse outcomes: mortality and major non-fatal outcomes [11]. Observational studies have an important role in the identification of large adverse effects of treatment on infrequent outcomes (i.e., rare but significant side effects). Such studies can also provide useful information about risks and help generalize from clinical trials to clinical practice.

Current hierarchical evaluations of evidence based on determining efficacy are also poorly attuned to the evaluation of diagnostic tests. Evaluation of the clinical relevance of a diagnostic test (such as GH stimulation tests) will depend on a number of factors including (1) the availability of an independent blind comparison with an adequate reference standard (and should this be height velocity or a 'well-established' test such as insulin hypoglycaemia?); (2) whether the study is based on patients with an appropriate spectrum of disease and those who are normal, and (3) the test sensitivity and specificity and the prevalence of the disorder under investigation.

Further confounding variables in the case of GH stimulation tests include whether there has been a recent spontaneous pulse of GH and the nature of the assay (RIA, IRMA/monoclonal, polyclonal, etc.) and the GH molecular types that are being recognized. Diagnostically, arbitrary cut-off values for GH 'sufficiency/insufficiency' are still used based either on historical data from the pituitary GH era (when GH therapy had to be reserved for those with severe GH deficiency and who needed and would potentially benefit from it most) or data obtained from different assay types – or both.

How many tests should be carried out for the diagnosis of GH 'deficiency'? Many centres, on the basis of NICE guidance, perform two GH stimulation tests and accept a single normal GH peak as sufficient to exclude GH deficiency. The 'rationale' for doing a second GH stimulation test and interpreting the results together is that a single test could often fail to achieve a normal response (for example because of a spontaneous GH pulse immediately prior to the test) and therefore a short child could be falsely labelled as GH-deficient.

However, tests are discordant in approximately 40–50% of cases with increasing discordance at more 'borderline' or equivocal GH levels. Requiring both tests to be positive for GH deficiency (GH response less than an arbitrary level) maximizes specificity (i.e. avoids falsely labelling normal children as GHD) but probably misses many 'treatable' individuals. Requiring both to be negative (response greater than an arbitrary GH level) maximizes sensitivity (i.e. minimizes missed diagnoses) but may falsely label many normal children as being GH-deficient. Conversely, if one abnormal test is deemed sufficient for the diagnosis of GH deficiency, around 50% of those diagnosed 'GH-deficient' may be diagnosed incorrectly [12].

The effect of applying different peak GH values in response to an ITT for the diagnosis of GH insufficiency using height velocity as the 'gold standard' demonstrates varying sensitivities and specificities despite comparable test efficiencies at all levels. Many diagnostic tests give the most discriminating answers when that is least necessary and least clear answers when there is clinical uncertainty also – auxological data remain important factors in diagnosing GH deficiency [13].

We still suffer from an overly 'black and white' view of diagnosis and disease [1, 14]. Following in the footsteps of Socrates (see below) and Wordsworth ('...that false secondary power/By which we multiply distinctions, then/Deem that our puny boundaries are things/That we perceive, and not that we have made'), Richard Dawkins has pointed out the inappropriateness of 'discontinuous minds' in the field of biology – 'We should not be in the business of drawing lines ... there is no law of nature which says that boundaries have to be clear cut' [15]. How we view the world scientifically is still overdependent on the 19th century 'Victorian' need to label, classify and pigeonhole as an aid to understanding. A 21st century view should be that classification can obscure understanding when there is limited knowledge: 'He who first gave names, gave them according to his conception of the things which they signified, and if his conception was erroneous, shall we not be deceived by him?' (Socrates quoted in Asher 1954) [16]. We can agree, with Humpty Dumpty [17], that 'GH deficiency' means (practically) whatever we want it to mean (" 'When *I* use a word" Humpty Dumpty said, in a rather scornful tone, "it means just what I choose it to mean, neither more nor less' ").

A Classification of paediatric endocrine disorders has recently been published [18] but the editors recognize that this will need to evolve as increasing understanding of aetiology and pathophysiology modify currently descriptive areas of classification.

A further problem is that we assume that because we can measure something (such as a serum concentration of GH or IGF-1) – or think we can! – these are the only or most important and clinically relevant things to assess – not everything that counts can be counted; not everything that can be counted counts [14]. Nevertheless, we are starting to recognize that, for example, stature is meaningless other than in comparison with others and that what matters to a short child is how (s)he feels about him/herself and how (s)he functions in society [14].

Doing more tests does not necessarily equate with better medicine. A reductionist scientific model – understanding systems by reducing them to their component parts – is based on linearity whereas the natural world is complex and depends on the interconnections of its components. Individual patients function at different biological, psychological and social levels with interaction and feedback between genes, cells, organs, organ systems, the whole body and the environment [19] – how things connect is important and evidence is not 'truth' [20]. Systems biology is reasserting the primacy of the whole organism. Identifying individual molecular abnormalities in particular clinical conditions is not an end in itself – systems biologists are studying the system properties of entire networks and it is how genes interact that matters [19, 21].

Evidence-based guidelines have been characterized as a reductionist approach to provide best advice about managing conditions [22]. One problem with a hierarchy of evidence 'strength' and thus of the strength of recommendations (as opposed to 'key' recommendations which, if implemented, would lead to greatest gain in health outcomes) is that governments, commissioners and health insurers are attracted to the idea of allocating 'cost-effective' health resources on the basis of 'evidence'. However, the often large quantities of trial data required to meet the standards of evidence simply do not exist. Evidence-based medicine can, therefore, introduce a systematic bias towards treatments in areas where there are existing funds available to show effectiveness (e.g., a pharmacological agent such as GH), and where there are more likely to be cost-effectiveness data available or easier to model, at the expense of other areas where rigorous evidence does not exist or is not easily attainable (e.g., psychological support) [23].

For a treatment to be not just 'effective' but cost-effective, its costs must be considered to be justified by the benefits provided compared to an existing treatment. A cost-effectiveness analysis will express results as a ratio of cost per unit of health outcome, with the latter normally expressed in 'natural units', e.g. symptom-free days, cases averted, life-years saved. There are thresholds in common use as ceiling values of cost-effectiveness ratios beyond which interventions are no longer considered cost-effective, reflecting the (arbitrary) maximum value which decision-makers attach to health benefits (commonly expressed in terms of a cost per quality-adjusted life year (QALY) gained – a QALY measures benefit as a combination of the impact of both length and QoL available from a treatment). Nevertheless, analyses of data in heart disease and cancer (derived from the Quest for Quality and Improved Performance research initiative of The Health Foundation in England), support the view that, with regard to QALYs, extra spending can give rise to distinctly better health outcomes if targeted in the right way [24]. This has implications for currently accepted cost-effectiveness thresholds as a basis for recommending adoption of a new technology or drug. In any case, defining such 'outcomes' and 'benefits' when GH is given to short stature children is not straightforward.

Medical decision-making is a complex process. If a patient has several diagnoses, each with several possible treatments and additionally there are symptoms and signs to consider as well as uncertain differential diagnoses, the complexity of the 'best practice' solution may be insoluble even using modern computer technology [22].

GH Therapy – How Do We Define Benefit and What Is the Evidence?

Nevertheless, an evidence-based approach to thinking about a clinical problem will involve (1) framing a clinically important and relevant question: in a particular patient or population to be studied (e.g., girls with Turner syndrome (TS)), will a particular intervention or exposure (e.g., GH) perhaps in comparison with another intervention (GH + oxandrolone) result in a greater increase in final height (an outcome of interest); (2) comprehensive review of available literature (using electronic databases, etc.) followed by critical appraisal of that literature, and (3) a decision as to the relevance to your own clinical practice (integration of external evidence with clinician and patient/parent views and values).

Applying evidence to inform high-quality patient care is not straightforward. For example, other methodological difficulties with many studies of childhood growth include the use of predicted or projected height comparisons, historical control groups, selection bias and variable treatment protocols. All too often, growth studies are still reported based on 'predicted adult height' or other surrogate outcomes (even final height is not a validated proxy for psychological contentment or QoL).

High-quality trials may demonstrate positive or negative efficacy outcomes but there will likely be publication bias as it is easier to publish a trial demonstrating a positive benefit than one demonstrating no effect. Poor quality or underpowered published trials may at best be uninformative but at worst may be seriously misleading. Historically, in the area of GH therapy many studies have been performed in too few patients to be certain about a positive effect or that a finding of a lack of effect really means that there is no effect.

For evaluating efficacy, clinical researchers are now more aware of the importance of designing adequately powered studies (national and international cooperation is often necessary) and having appropriate control groups (placebo and double-blinded where ethically and practically possible). However, the decision to prescribe a drug is based on a balance of benefit and risk, and the medical profession (like the general public) are not good at assessing risks – the likelihood of a particular diagnosis (like the sensitivity and specificity of a diagnostic test itself) depends not only on the patient's symptoms and signs but also on the incidence and prevalence of the disease in the relevant population.

The way data are presented can also have a significant impact on the decisions we make as prescribers or patients. Healthcare practitioners or consumers are much more likely to choose an intervention whose benefit is expressed as a relative risk reduction [25]. If the same risks from the same trial are expressed as an absolute risk reduction, they generally look less impressive because absolute risk levels in populations studied are often very low. The most appropriate way of expressing benefit from an intervention is to express the risk in terms of numbers needed to treat to produce that benefit. This provides information on the potential value of an intervention in both effectiveness and cost-effectiveness terms in a way that can be understood intuitively as relevant to healthcare practice.

A comprehensive systematic review of the relevant scientific literature may already have been carried out and, if so, will save the clinician considerable time and effort. These are much less biased than traditional (narrative) reviews produced by 'experts' in the field. However, systematic reviews can also be badly conducted and have significant methodological flaws, not least due to potential conflicts of interest. Some meta-analyses dramatically clarify evidence, but others are scientifically flawed by extrapolating evidence from heterogeneous populations in RCTs to one-to-one consultations with individuals [20].

Several Cochrane reviews have looked at the evidence base behind GH therapy in various clinical contexts. In TS, a 2003 Cochrane Review [26] found only 4 RCTs that included 211 participants after 1 year of treatment of sufficient quality for inclusion of 1 year treatment data (described in six publications!). Three studies were included in the analyses of growth outcomes (one study did not report any data) and only one trial reported adult height outcomes. The reviewers concluded that recombinant human GH in doses between 0.3 and 0.375 mg/kg/week increased short-term growth in girls with TS by approximately 3 cm in the first year of treatment and by approximately 2 cm/year after 2 years of treatment. At that time there was little evidence on the effects of GH on adult height but in the one trial reporting these data this increased by approximately 5 cm over an untreated control group – still >2 SD below the normal height range.

They recommended that additional trials of the effects of GH carried out with control groups until final height is achieved would allow better informed decisions about whether the benefits of GH treatment outweigh the requirement of treatment over several years at considerable cost. By then, use of GH to treat TS girls in clinical practice was so widespread that it was difficult or impossible to obtain ethical committee consent for placebo-controlled GH therapy studies and studies continued to suffer from a number of methodological defects including predicted or projected height data, selection bias, variable treatment protocols and historical control groups.

By the time of a repeat review in 2007 [27], there were data from 365 patients in four trials and still only one trial to final height. The increase in final height had risen by 1 cm and the call for further properly controlled studies to final height to better inform clinical practice was repeated. It took nearly 20 years for the first well-conducted RCT in the use of GH in TS to be reported [28]. Even here there are problems however: a third of the patients withdrew from the study (was this because GH was so effective/ineffective/painful, etc.?) and there are no data on starting GH therapy at younger than 7 years. The mean height gain due to GH was found to be 7.3 cm (but with wide confidence intervals: CI 5.4 cm, 9.2 cm). Starting GH at 7–8 years of age was more effective than at 12–13 years but there are no data on starting earlier (a 'reasonable' objective if the diagnosis is made early enough?). There was a highly variable individual response to therapy but puberty could be induced at a 'near physiological' age (13 years) without disadvantage in height outcome. There were more ENT disorders (tonsil and adenoid hypertrophy) and more joint disorders.

This still leaves the clinician who wishes to practice 'evidence-based medicine' in some difficulty therefore: height is not a validated proxy for psychological wellbeing or QoL and, based on French population TS data (questioning of adult women with TS), two thirds of the TS patients treated with GH in the Canadian study would not have considered the treatment to have been worthwhile based on these results [29]. However, it is likely that the preoccupations and priorities of TS patients change with age: preschool children are unconcerned by their height and cannot see a rationale for daily injections of GH; older prepubertal girls are aware of their height and (over)optimistic about the effects of GH; from early puberty many are more realistic about their height prognosis and self-conscious about their immature body image (which is helped by oestrogen therapy); adults are more concerned by infertility than their height.

At what age should treatment start? Is it 'common sense' or mere supposition that trying to keep these girls similar in height to their peers throughout as much of childhood as possible would be beneficial psychologically? How is the high-quality evidence about height gain to be put in perspective against the lack of long-term safety evidence (otitis media, colon cancer, metabolic) and the lack of evidence for psychosocial benefit?

A Cochrane review of the use of GH for children with chronic renal failure in 2001 [30] identified 10 RCTs of sufficient quality on 481 children and found that, on average, 1 year of 'pharmacological' GH doses (28 IU/m²/week) resulted in a 4-cm/year increase in height velocity above that of untreated controls. However, it was still uncertain if this would result in an increase in adult height and it was suggested that the benefits of longer courses or higher doses of treatment warranted further study, although a further review by the same authors later that year [31], concluded that there was no demonstrable benefit from longer courses or higher GH doses.

By the time of the updated Cochrane Systematic Review in 2006 [32], there were 15 RCTs involving 629 children identified – the height velocity increase over untreated controls was marginally lower at 3.8 cm/year but it was concluded that the trials were too short to determine if continuing treatment resulted in an increase in adult height.

A particularly controversial area for GH prescribing has been in so-called 'idiopathic short stature' (ISS). Whilst GH therapy for clearly GH-deficient short children is well accepted, treatment of those with no recognizable abnormalities in their GH–IGF1 axis or other recognized pathology remains controversial. However, as discussed above and in Chapter 1, defining GH deficiency is arbitrary due to the poor efficiency, sensitivity and specificity of GH stimulation tests [33], and many patients labelled as having idiopathic GHD, treated successfully with GH, would be better categorized as ISS. Likewise, many currently diagnosed as GH-deficient from GH stimulation tests may have normal spontaneous GH secretion and IGF-1 concentrations and should be diagnosed as ISS [34, 35]. Some children may have high GH levels but remain short because of GH insensitivity. We remain poor at predicting response and evaluating appropriate endpoints for treatment efficacy – final height, short-term catch-up growth, QoL, metabolic parameters – both in classical GH deficiency and in other disorders where GH therapy seems 'effective' such as TS or chronic renal failure [36].

'Normal' means physiologically correct, not 'average', 'common' or 'conventional' [14]. The World Health Organization defines health as not merely absence of disease, but a complete state of mental, physical and social wellbeing. Moral objections to growth-promoting treatment may be based on the view that the aim of treatment is to enhance normal physical characteristics, but a crucial purpose in increasing adult height (if achievable) would be resulting psychological or QoL gain [37, 38]. In the context of other therapies for conditions not related to disease or dysfunctional states, treatment of short children without 'GH deficiency' to normalize conventionally defined abnormally short stature should be seen as ethically acceptable. However, once confounding factors (e.g., socio-economic level) have been excluded, children with short stature have not been shown, as a group, to display clinically significant behavioural or emotional problems [39, 40]. On an individual basis, however, short stature may be the additional stressor which leads to psychological maladaptation.

Cochrane reviewers have considered the ISS evidence systematically. In 2003 [41], 9 RCTs were included but only one reported even near adult height in girls, who were 7.5 cm taller than untreated controls All other trials reported only short-term outcomes. Results suggested that short-term height gains can range from nil to approximately 0.7 SD over 1 year, which makes prediction for individual children very difficult. One study reported health-related QoL and showed no significant improvement in GH-treated children compared with controls. No serious adverse treatment effects were reported. It was concluded that therapy can increase short-term growth and improve (near) adult height but increases are such that treated individuals remain relatively short compared with peers. Further research in the form of large, multicentre RCTs were recommended which should focus on adult height and address QoL and cost issues.

By 2007, the same authors [42] found more but still a paucity of adult height data. One further trial had found that children treated with GH were 3.7 cm taller than children in a placebo-treated group; other trials still only reported short-term outcomes. One study found no significant evidence that GH treatment improves psychological adaptation or self-perception in children with ISS. The recommendations were unchanged from 4 years previously.

Recent consensus-derived recommendations for GH therapy on ISS have been published and their basis set out in greater detail in two further publications [43–45]. ISS was defined auxologically as a height below –2 SD without evidence of disease following a complete evaluation (including stimulated GH levels) by a paediatric endocrinologist. It was concluded that whilst ISS may be a risk factor for psychosocial problems, true psychopathology is rare. In the USA and 7 other countries, regulatory authorities have approved GH treatment (at doses up to 53 µg/kg/day) for children shorter than –2.25 SDS, whereas in other countries lower cut-offs are proposed, and the licensing of GH therapy for this indication in Europe seems unlikely. Importantly, it emphasized that it is worthwhile to consider psychological counselling instead of, or as an adjunct to, GH treatment and reiterated that predicted height may be inaccurate and is not an absolute criterion for GH treatment. Nevertheless, on an individual basis, the shorter the child, the more consideration should be given to GH therapy and the mean increase in adult height in children with ISS attributable to GH therapy (average duration of 4–7 years) is 3.5–7.5 cm but with highly variable individual responses. GH therapy for children with ISS appears to have a similar safety profile to other GH indications.

In this potential therapeutic area, individual prescribing decisions must be made against a background of systematic evidence review and explicit linkage between recommendations and graded evidence levels. Statistically significant effects (e.g., increased final height) must be evaluated in terms of clinical benefit – the presence and potential for alleviation of short stature-related suffering is crucial [46].

How useful is growth screening, and is growth monitoring in children over time more valuable in identifying children with underlying pathology and/or who might benefit from GH therapy? A Cochrane review published in 2002 [47] and a previously published analysis by the same authors [48] identified only two randomized trials, both conducted in developing countries, comparing routine growth monitoring combined with intervention when growth was abnormal, with no growth monitoring. The review concluded that there was 'insufficient reliable information to be confident as to whether routine growth monitoring is of benefit to child health in developing and developed country settings' and that 'given the level of investment in growth monitoring worldwide, it is surprising that there is so little research evaluating its potential benefits and harms'.

Height Screening/Monitoring

Height screening to detect disease has been the focus of a number of studies and several of these have reported the presence of organic disease in a proportion of children with short stature identified in screening programs. Both the Wessex and Oxford growth studies of pre-school children identified a number of children with short stature of whom 17–21% had underlying organic disease, including GH deficiency and TS [49, 50], suggesting that screening at school entry would detect a number of children with underlying pathology. There are few studies in school age children however, although the Utah growth study of 114,881 children in elementary school identified 40 children with short stature subsequently found to have an underlying endocrine disorder, including 16 with previously unrecognized GH deficiency and 6 girls with TS [51]. Van den Broeck et al. [52] found a small number of previously unrecognized cases of GH deficiency and TS in 5- to 12-year-old children and height velocity to be a useful screening tool in this age group. Van Buuren et al. [53] reported that growth monitoring was useful in the diagnosis of TS, finding that a rule taking into account absolute height SDS, parental height and deflection in height velocity gave a specificity of 99.4% and a sensitivity of 79% for the identification of TS. Clearly, such studies are also needed in other conditions.

Current recommendations in the UK are based on the suggestion that routine growth monitoring to detect centile crossing has too low a sensitivity and specificity to be regarded as screening [54]. Health for All Children [55] recommends a minimum of weight measurement at 2, 3, 4, 8 and 12–15 months, and at 3–4 years, with height and weight at school entry. Routine growth monitoring is not recommended after the age of 5 years. Additionally, the recommendations are that 'assessment of growth should be part of the minimum evaluation of any child presenting with chronic poor health, atypical development, other unexplained symptoms, or concern about their psychological and social wellbeing'.

There is, however, considerable ongoing debate about how often measurements should be made in the school age children to identify those in whom growth falters in mid-childhood (e.g., acquired hypothyroidism) and to detect those whose growth disorder has not been identified by school entry, including a significant proportion of children with GH deficiency and girls with TS. In particular, the British Society for Paediatric Endocrinology and Diabetes (BSPED) has expressed concerns about the recommendations (www.BSPED.org.uk). While endorsing 'universal measurements for all children, which should be made at the very minimum on one occasion, possibly combined with school entry assessment', it acknowledges that there is 'no clear scientific evidence as to the most efficient schedule'. It urges that 'scientific evidence needs to be accumulated with the proposed introduction of a single measurement at school entry policy to determine whether this is indeed the most cost-effective and efficient approach in comparison with more frequent or no measurements being taken'. Additionally, it reaffirms its previous statement that whenever practical, a growth assessment should be made at every point of contact with a health professional as this is good clinical practice.

In the Netherlands, auxological referral criteria, as then defined, resulted in too many unnecessary referrals [56] but guidelines and an algorithm have now been proposed which show a high sensitivity with an acceptably low false-positive rate in 3- to 10-year-olds [57]. The authors emphasize that that no algorithm can fully replace clinical judgment (which takes into account information from the medical history and clinical examination).

Bone Age and Predicted Height as a Surrogate

Predictions of adult height incorporate information about skeletal maturity (bone age). An estimate of bone age (BA) defines the proportion of growth which has already taken place and the amount to come. BA standards were developed in normal children, and were only ever intended to be an indicator of skeletal maturation in healthy children although they are used most frequently to assess children who fall outside the design limits, e.g. those with pathology or children on growth-stimulating (e.g., GH) or -suppressing (e.g. glucocorticoid) medication.

Whilst BA assessment is a key element in adult height prediction, BA standards are cross-sectional and based on normal children of 'average' height reaching puberty at an average age and progressing through it at an average tempo. In practice, BA estimations on the same individual repeated through puberty do not show increments of 1 'year' BA per year of chronological age, because BA accelerates and deviates from a cross-sectional-based centile in a way similar to that for height. Thus, children with delayed puberty may manifest an increasingly retarded BA (as the standards are based on those with puberty at an average age), and then show a rapid BA 'catch-up' in puberty. Conversely, in those entering puberty at an age younger than average, rapid acceleration and progressive advancement of BA occurs.

Thus predicted adult height is problematic in children with pathology affecting growth [58] and a delayed BA at the start of treatment gives misleadingly high height predictions and treatment effects will be underestimated. Conversely, studies of (for example) gonadotrophin-releasing hormone analogues purporting to show benefit compared to predicted height in children with precocious puberty may significantly overestimate treatment benefit. Target height can also be misleading (does one or other parent have the same pathological condition as the child?) and, if population heights are not taken into account, adult height is underestimated and the treatment effect overestimated.

Where Does Evidence-Based Child Health Fit in Endocrinology

EBCH is a useful basis on which to practice the 'science' (as opposed to the equally important 'art') of medicine. However, there has developed a healthy debate about the role and relevance of evidence-based medicine for the practising clinician [59–64]. As Miles et al. [64] have stated: 'when taken to extremes by its diehard protagonists, evidence-based medicine can, in fact, be described both as unscientific and antiscientific. Unscientific because it is based on assumption rather than a proven hypothesis that it can improve both the process and outcome of medical care. Antiscientific because it has favoured its own internal top-down approach to the identification of 'knowledge for practice'.

In reality, not all questions of relevance to clinical decision-making are scientific in their nature. Medicine is at the interface between biology, sociology and psychology and clinical decisions are typically made through a plurality of means formulated on 'the evidence of the clinic' (clinical experience, etc.), and not simply through scientific data derived from RCTs, meta-analyses and systematic reviews [60].

It has even been said that quantitative data derived from methodologically limited studies conducted on unrepresentative populations in highly controlled experimental environments cannot in themselves form an adequate basis for clinical decision-making in a clinical consultation between an individual doctor and patient [60]. Outcomes in individual patients may be very different from mean responses in clinical trials – sometimes better but often (much) worse.

In fact, observational studies and RCTs contribute complementary clinical information [11]. Wider appreciation of the different strengths and weaknesses of different types of study design would increase the likelihood that the most reliable evidence available informs clinical decisions about treatments that doctors use for a wide range of conditions.

Conclusions

Short stature is not a disease and normal is not the same as average. Most children referred with short stature have no evidence of underlying pathology and will have genetic short stature and/or constitutional growth delay. Many studies dealing with the psychosocial consequences of short stature are based on children referred for investigation, who are therefore more likely to be those with other pathology or to have greater problems with psychosocial adaptation (or parents who perceive that they will do so). Evidence suggests that there is little effect from being shorter than average on psychological functioning in children, adolescents or adults.

Large-scale prospective studies are needed in order to determine the best growth monitoring strategy for children. GH deficiency is primarily a clinical diagnosis, supported by auxological, biochemical and radiological findings and, ultimately, the growth response to GH therapy is the best clinical marker for GH deficiency. The continued search for the 'Holy Grail' of the perfectly sensitive and specific test to define GH deficiency will prove fruitless in the context of the continuum of GH secretory ability and sensitivity in children of different heights.

Studies on the pharmacological use of GH in different short stature conditions must be designed not just to determine the most effective strategy to improve adult height, but also the balance between any effects of height gained on physical health and social wellbeing against effects on QoL of daily injections, potential short- and long-term adverse effects and treatment costs. There should be ongoing surveillance of GH therapy for both licensed and unlicensed indications including long-term efficacy and safety data (including safety data in adults treated with GH in childhood only).

Psychological and QoL outcomes should be central to the evaluation of treatment benefit in any future studies, or in decision-making meanwhile by the clinician. Unless broader questions are raised as to what constitutes 'benefit', the debate in treating short children will remain at the 'centimetres gained' (or even 'predicted to be gained') level, and properly designed studies to address important wider safety issues and psychological and QoL outcomes will not take place.

Doctors are unique in their ability to generate differential diagnoses, a clinical diagnosis and the exercise of considered judgment in initiating management and treatment most appropriate to an individual patient's needs. As an appropriately and rigorously developed tool, EBCH should have a significant role to play in improving patient care but there also remains a challenge to demonstrate (rather than assume) that evidence-based medicine can improve both the process and outcome of medical care.

References

1   Kelnar CJH: Evidence-based child health – SIGN and NICE. Arch Dis Child Educ Pract Ed 2008; 93:190–198.

2   Moyer V, Elliott E: Preface; in Moyer V, Elliott E, Davis R, et al (eds): Evidence-Based Paediatrics and Child Health. London, BMJ Books, 2000.

3   Stylianou C, Kelnar CJH: The Introduction of Successful Insulin Treatment for Diabetes [commentary on Banting FG, Best CH, Collip JB, Campell WR, Fletcher AA: Pancreatic extracts in the treatment of diabetes mellitus. CMAJ 1922; 12:141–146]. The James Lind Library (www.jameslindlibrary.org).

4   Collins R, MacMahon S: Reliable assessment of the effects of treatment on mortality and major morbidity. I. Clinical trials. Lancet 2001;357:373–380.

5   Tanner JM: Fetus into Man. Boston, Harvard University Press, 1978.

6   Centers for Disease Control (CDC): Fatal degenerative neurologic disease in patients who received pituitary-derived human growth hormone. MMWR Morb Mortal Wkly Rep 1985;34:359–60, 365–366.

7   Powell-Jackson J, Weller RO, Kennedy P, Preece MA, Whitcombe EM, Newsom-Davis J: Creutzfeldt-Jakob disease after administration of human growth hormone Lancet 1985;ii:244–246.

8   Gibbs CJ Jr, Joy A, Heffner R, Franko M, Miyazaki M, Asher DM, Parisi JE, Brown PW, Gajdusek DC: Clinical and pathological features and laboratory confirmation of Creutzfeldt-Jakob disease in a recipient of pituitary-derived human growth hormone N Engl J Med 1985;313:734–738.

9   Brown P, Gajdusek DC, Gibbs CJ Jr, Asher DM: Potential epidemic of Creutzfeldt-Jakob disease from human growth hormone therapy N Engl J Med 1985;313:728–731.

10 Wilton P: Adverse events reported in KIGS; in Ranke MB, Price DA, Reiter EO (eds): Growth Hormone Therapy in Paediatrics – 20 Years of KIGS. Basel, Karger, 2007, pp 432–441.

11 MacMahon S, Collins R: Reliable assessment of the effects of treatment on mortality and major morbidity. II. Observational studies. Lancet 2001; 357:455–456.

12 Drake AM, Kelnar CJH: The evaluation of growth and the identification of growth hormone deficiency. Arch Dis Child 2006;91:ep61–ep67.

13 Obara-Moszyńska M, Kedzia A, Korman E, Niedziela M: Usefulness of growth hormone (GH) stimulation tests and IGF-I concentration measurement in GH deficiency diagnosis. J Pediatr Endocrinol Metab 2008;21:569–579.

14 Kelnar CJH: Pride and prejudice – stature in perspective. Acta Paediatr Scand 1990;370(suppl):5–15.

15 Dawkins R: Gaps in the mind; in A Devil's Chaplain – Selected Essays. London, Weidenfeld & Nicolson, 2003.

16 Asher R: Straight and crooked thinking in medicine. BMJ 1954;ii:460–462.

17 Carroll L (Charles Lutwidge Dodgson): Through the Looking-Glass, and What Alice Found There. 1871.

18 Wit J-M, Ranke MB, Kelnar CJH (eds): ESPE Classification of Paediatric Endocrine Diagnoses. Horm Res Spec Ed 2007;68(suppl 2):1–120.

19 Noble D: The Music of Life – Biology beyond Genes. Oxford, Oxford University Press, 2006.

20 Sweeney K, Kernick D: Clinical evaluation: constructing a new model for post-normal medicine. J Eval Clin Pract 2002;8:131–138.

21 Dawkins R: Hierarchical organisation: a candidate principle for ethology; in Bateson PPG, Hinde RA (eds): Growing Points in Ethology. Cambridge University Press, 1976, pp 7–54.

22 Welsby PD: Evidence-based medicine, guidelines, personality types, relatives and absolutes. J Eval Clin Pract 2002;8:163–166.

23 Georgiou A: Data information and knowledge: the health informatics model and its role in evidence-based medicine. J Eval Clin Pract 2002;8:127–130.

24 Martin S, Rice N, Smith PC: The Link between Healthcare Spending and Health Outcomes – Evidence from English Programme Budgeting Data. London, The Health Foundation, 2007.

25 Malenka DJ, Baron JA, Johansen S, Wahrenberger JW, Ross JM: The framing effect of relative and absolute risk. J Gen Intern Med 1993;8:543–548.

26 Cave CB, Bryant J, Milne R: Recombinant growth hormone in children and adolescents with Turner syndrome. Cochrane Database Syst Rev CD003887, 2003.

27 Baxter L, Bryant J, Cave CB, Milne R: Recombinant growth hormone for children and adolescents with Turner syndrome. Cochrane Database Syst Rev CD003887, 2007.

28 Stephure DK, Canadian Growth Hormone Advisory Committee: Impact of growth hormone supplementation on adult height in Turner syndrome: results of the Canadian randomized controlled trial J Clin Endocrinol Metab 2005;90:3360–3366.

29 Carel JC: Growth hormone in Turner syndrome: twenty years after, what can we tell our patients? J Clin Endocrinol Metab 2005;90:3930–3934.

30 Vimalachandra D, Craig JC, Cowell C, Knight JF: Growth hormone for children with chronic renal failure. Cochrane Database Syst Rev CD003264, 2001.

31 Vimalachandra D, Craig JC, Cowell CT, Knight JF: Growth hormone treatment in children with chronic renal failure: a meta-analysis of randomized controlled trials. J Pediatr 2001;139:560–567.

32 Vimalachandra D, Hodson EM, Willis NS, Craig JC, Cowell C, Knight JF: Growth hormone for children with chronic kidney disease. Cochrane Database Syst Rev CD003264, 2006.

33 Hindmarsh PC: Endocrine assessment and principles of endocrine testing; in Kelnar CJH, Savage MO, Saenger P, Cowell CT (eds): Growth Disorders, ed 2. London, Hodder Arnold, 2007, pp 219–228.

34 Hailey JA, Bath LE, Kelnar CJH: Idiopathic short stature – diagnostic and therapeutic dilemmas. Royal Society of Medicine Current Medical Literature – Growth Hormone and Growth Factors 1999;14:61–65.

35 Kelnar CJH, Albertsson-Wikland K, Hintz RL, Ranke MB, Rosenfeld RG: Should we treat children with idiopathic short stature? Horm Res 1999;52:150–157.

36 Juul A, Bernasconi S, Chatelain P, Hindmarsh P, Hochberg Z, Hokken-Koelega A, de Muinck Keizer-Schrama SMPF, Kiess W, Oberfield S, Parks J, Strasburger CJ, Volta C, Westphal O, Skakkebaek NE: Diagnosis of growth hormone (GH) deficiency and the use of GH in children with growth disorders. Horm Res 1999;51:284–299.

37 Downie AB, Mulligan J, Stratford RJ, Betts PR, Voss LD: Are short normal children at a disadvantage? The Wessex Growth Study. BMJ 1997;314:97–100.

38 Rekers-Mombarg LT, Busschbach JJ, Massa GG, Dicke J, Wit JM: Quality of life of young adults with idiopathic short stature: effect of growth hormone treatment. Dutch Growth Hormone Working Group. Acta Paediatr 1998;87:865–870.

39 Sandberg DE, Brook AE, Campos SP: Short stature: a psychosocial burden requiring GH therapy? Pediatrics 1994;94:832–840.

40 Zimet GD, Owens R, Dahms W, Cutler M, Litvene M, Cuttler L: Psychosocial outcome of children evaluated for short stature. Arch Pediatr Adolesc Med 1997;151:1017–1023.

41 Bryant J, Cave C, Milne R: Recombinant growth hormone for idiopathic short stature in children and adolescents. Cochrane Database Syst Rev CD004440, 2003.

42 Bryant J, Baxter L, Cave CB, Milne R: Recombinant growth hormone for idiopathic short stature in children and adolescents. Cochrane Database Syst Rev CD004440, 2007.

43 Cohen P, Rogol AD, Deal CL, Saenger P, Reiter EO, Ross JL, Chernausek SD, Savage MO, Wit JM on behalf of the 2007 ISS Consensus Workshop Participants: Consensus Statement on the Diagnosis and Treatment of Children with Idiopathic Short Stature: A Summary of the Growth Hormone Research Society, the Lawson Wilkins Pediatric Endocrine Society, and the European Society for Paediatric Endocrinology Workshop. J Clin Endocrinol Metab 2008;93:4210–4217.

44 Wit JM, Clayton PE, Rogol AD, Savage MO, Saenger PH, Cohen P: Idiopathic short stature: definition, epidemiology, and diagnostic evaluation. Growth Horm IGF Res 2008;18:89–110.

45 Wit JM, Reiter EO, Ross JL, Saenger PH, Savage MO, Rogol AD, Cohen P: Idiopathic short stature: management and growth hormone treatment. Growth Horm IGF Res 2008;18:111–135.

46 Haverkamp F, Ranke MB: The ethical dilemma of growth hormone treatment of short stature: a scientific theoretical approach. Horm Res 1999;51:301–304.

47 Panpanich R, Garner P: Growth monitoring in children. Cochrane Database Syst Rev CD001443, 2000.

48 Garner P, Panpanich R, Logan S: Is routine growth monitoring effective? A systematic review of trials. Arch Dis Child 2000;82:197–201.

49 Voss LD, Mulligan J, Betts PR, Wilkin TJ: Poor growth in school entrants as an index of organic disease: the Wessex Growth Study. BMJ 1992;305:1400–1402.

50 Ahmed ML, Allen AD, Sharma A, Macfarlane JA, Dunger DB: Evaluation of a district growth screening programme: the Oxford Growth Study. Arch Dis Child 1993;69:361–365.

51 Lindsay R, Feldkamp M, Harris D, Robertson J, Rallison M: Utah Growth Study: growth standards and the prevalence of growth hormone deficiency. J Pediatr 1994;125:29–35.

52 Van den Broeck J, Hokken-Koelega A, Wit JM: Validity of height velocity as a diagnostic criterion for idiopathic growth hormone deficiency and Turner syndrome. Horm Res 1999;51:68–73.

53 Van Buuren S, van Dommelen P, Zandwijken GR, Grote FK, Wit JM, Verkerk PH: Towards evidence-based referral criteria for growth monitoring. Arch Dis Child 2004;89:336–341.

54 Hall DMB: Growth monitoring. Arch Dis Child 2000;82:10–15.

55 Hall DMB, Elliman D: Health for All Children, ed 4. Oxford, Oxford University Press, 2003.

56 Van Buuren S, Bonnemaijer-Kerkhoffs DJ, Grote FK, et al: Many referrals under Dutch short stature guidelines Arch Dis Child 2004;89:351–352.

57 Grote FK, van Dommelen P, Oostdijk W, et al: Developing evidence-based short stature guidelines for referral for short stature. Arch Dis Child 2008;93:212–217.

58 Kelnar CJH, Stanhope R Height prognosis in girls with central precocious puberty treated with GnRH analogues. Clin Endocrinol 2002;56:295–296.

59 Miles A, Bentley P, Polychronis A, Grey JE, Melchiorri C: Recent developments in the evidence-based healthcare debate. J Eval Clin Pract 2001;7:85–89.

60 Miles A, Grey JE, Polychronis A, Melchiorri C: Critical advances in the evaluation and development of clinical care. J Eval Clin Pract 2002;8:87–102.

61 Miles A, Grey JE, Polychronis A, Price N, Melchiorri C: Current thinking in the evidence-based health care debate. J Eval Clin Pract 2003;9:95–109.

62 Miles A, Grey JE, Polychronis A, Price N, Melchiorri C: Developments in the evidence-based health care debate – 2004. J Eval Clin Pract 2004;10:129–142.

63 Miles A, Loughlin M: Continuing the evidence-based health care debate in 2006. The progress and price of EBM. J Eval Clin Pract 2006;12:385–398.

64 Miles A, Loughlin M, Polychronis A: Medicine and evidence: knowledge and action in clinical practice. J Eval Clin Pract 2007;13:481–503.

Christopher J.H. Kelnar, MD
Department of Paediatric Endocrinology, Section of Child Life and Health
Division of Reproductive and Developmental Sciences, University of Edinburgh
20 Sylvan Place, Edinburgh EH9 1UW (UK)
Tel. +44 0131 536 0611, Fax +44 131 536 0821, E-Mail chris@kelnar.com

Hindmarsh PC (ed): Current Indications for Growth Hormone Therapy, ed 2, revised.
Endocr Dev. Basel, Karger, 2010, vol 18, pp 40–54

Safety of Recombinant Human Growth Hormone

Jean-Claude Carel[a] · Gary Butler[b]

[a]Pediatric Endocrinology and Diabetology, and INSERM U690, University Paris 7, Denis Diderot, Hôpital
Robert Debré, Paris, France, and [b]University College London Hospital and UCL Institute of Child Health,
London, UK

Abstract

With an increasing spectrum of indications for growth hormone (GH), knowledge of the short-
and long-term safety of this treatment is essential. In this chapter we review the main adverse
effects that have been demonstrated or discussed after long-term GH treatment in children. It
is well recognized that plasma insulin concentrations increase during GH treatment. The incidence of type 2 diabetes is raised during GH treatment, especially in subjects with other risk
factors. Recommendations include assessing glucose tolerance by measuring plasma glucose
and HbA1c before and during treatment. There is no consensus on the indications for insulin
measurement and/or oral glucose tolerance tests. Other recognized short-term complications
of GH treatment include pseudo-tumour cerebri, otitis media (Turner syndrome), and orthopaedic problems such as worsening of scoliosis and slipped femoral epiphysis. There are
reports of sudden death within the first 6 months of treatment in children with Prader-Willi
syndrome, mostly associated with severe obesity. Large cohort follow-up studies suggest that
children treated with GH following childhood cancer treatment do not have a greater number
of relapses, but there may be a higher incidence of second primary tumours in the early years
of GH therapy. A cohort treated with human pituitary GH showed a higher incidence of
tumours, and a GH effect on tumorigenesis has been seen in follow-up studies of acromegaly
raising the question of whether de novo cancer risk may be increased. A new prospective
pan-European safety surveillance study (SAGhE) has been launched to address these essential
questions. The role of monitoring the IGF-1 response (total or free concentrations) to GH
treatment to predict long-term safety is unclear at present. Attempts to target IGF-1 levels
within the normal range may result in the use of excessive doses of GH. In general, higher
dosage GH regimens may be associated with supraphysiological IGF-1 levels.

The indications for growth hormone (GH) treatment have evolved progressively
from the simple hormonal substitution process in severe GH deficiency to the

pharmacological manipulation of growth in an increasing number of diagnoses of short stature without pituitary hormone deficiency. GH deficiency is itself a very heterogeneous condition ranging from severe pituitary deficiency to cases of short stature selected for failure to pass a benchmark level in GH stimulation tests.

In the context of an increasing spectrum of indications for GH, knowledge of the short- and long-term safety of this treatment is essential. The history of transmission of prion diseases by pituitary-derived GH has led paediatric endocrinologists and the pharmaceutical industry to initiate pharmacovigilance programmes on recombinant GH early on. There is abundant data on the safety profile of GH during treatment. This profile is globally considered as favourable, as reviewed in a consensus conference held on the topic published in 2001 [1]. However, the available data do not address the important question of long-term safety. In this review, we will summarize the follow-up points during treatment, the available data on long-term safety and the ongoing studies (tables 1–3).

Short-Term Safety Issues

Diabetes and Glucose Intolerance
We will not review the complex effect of the GH/IGF-1 axis on insulin secretion and sensitivity, but it is well recognized that plasma insulin concentrations increase during GH treatment. The KIGS study has shown an increased incidence of type 2 diabetes with 1 case for 2,900 patient-years of treatment, a sixfold increase relative to the general population [2]. This excess risk seems particularly focused on certain groups with a higher risk (Turner syndrome, obese patients). The incidence of type 1 diabetes is not increased. In practice, it is recommended to assess glucose tolerance by measuring plasma glucose and HbA1c before treatment and at 6-monthly intervals initially, thereafter annually. There is no consensus on the indication for more sophisticated modalities of monitoring such as insulin measurement and/or oral glucose tolerance tests. Given the poor reproducibility of the oral glucose tolerance test, we recommend that it is only performed in selected cases such as at-risk individuals or in those with abnormal screening tests [3, 4].

Pseudo-Tumour Cerebri
Pseudo-tumour cerebri (benign intracranial hypertension) describes the intracranial fluid retention induced by initiation of treatment with GH and is observed in about 1 in 1,000 children treated [1 5]. It presents with headaches and papilloedema on fundoscopy. GH treatment should be stopped and, after recovery, can be reinstated by reducing the starting dose and then progressively increasing it over a few weeks to the full amount.

Table 1. Suggestions of key items of information which should be given to the family prior to their child starting GH treatment

GH safety information for parents and patients

Check-ups and routine safety monitoring
- Children on GH treatment should be reviewed by the specialist team every 3–6 months. The check-up should include a measurement of growth, general clinical examination and a review of development in puberty (where appropriate).
- A blood test should be taken at least once a year to measure growth factors (IGF-1), routine safety checks (table 2) and any other pituitary hormones deemed necessary by your doctor. An X-ray of the left hand and wrist is sometimes requested to follow the stages of bone growth and to predict adult height.
- It is not necessary to stop GH treatment when your child has another short illness.

Side effects of GH treatment
- The side effects of GH treatment are rare, thought to be around 1 in 1,000 children treated.
- Adverse reactions (pain and bruising) at injection sites are uncommon, but if they occur, advice from the specialist team will usually help to resolve them.
- Benign raised intracranial pressure (raised pressure in the fluid in the brain) may occur in the first few weeks of GH treatment. This may cause headaches, nausea and vomiting (rarely). Sometimes vision is blurred and examination of the eye can show papilloedema (a swelling of the optic disc). GH treatment is usually stopped for a short while and restarted slowly to prevent recurrence of the symptoms.
- Orthopaedic (bony) problems can include slipped femoral (hip) epiphysis which will usually need an orthopaedic operation, but GH treatment does not usually need to be stopped. Your child should be examined for the development of scoliosis (twist of the spine) especially during adolescence. This may be more common in Prader-Willi syndrome and in skeletal dysplasias.
- A small increase in blood glucose (sugar) and insulin levels may be seen in some children but this is reversible on stopping GH treatment.
- In girls with Turner syndrome there may be an increase in middle ear problems.
- Unusual other side effects include acute pancreatitis, prepubertal gynaecomastia (breast development), increase in the number and size of skin moles. Joint pains and carpal tunnel syndrome are more often seen in adult patients.
- The development of antibodies to GH treatment is very rare.
- There is no increased likelihood of developing leukaemia or any other cancer when starting GH in a healthy child. However, GH treatment is not started or continued in children who have active malignant disease. However, once the disease is under control or in remission, there is no evidence that GH treatment is associated with an increased risk of relapse.
- There is no risk of developing Creutzfeldt-Jakob disease from treatment with biosynthetic GH which has been used exclusively for the past 25 years.

In conclusion
- GH treatment is a safe and effective treatment but requires careful attention from everybody involved (children, parents, nurses, doctors) to ensure that the best outcome is achieved. Check-ups should continue at least until growth has been completed.
- For those patients who need to continue GH treatment into adult life, supervision should take place with a specialist in adult endocrinology.

Table 2. Checklist of safety controls before and during GH treatment (does not include the list of safety controls for specific conditions such as Turner syndrome)

- General clinical examination, with special emphasis on growth,body proportions and limbs
- Blood pressure
- Fasting plasma glucose
- Fasting insulin (requested in the SPC for children born SGA; clinical significance questionable)
- Biochemical bone profile
- HbA1c
- Thyroid function tests (in GH-deficient patients)
- IGF-1
- IGFBP-3 (requested in the SPC for children born SGA; clinical significance questionable)

Orthopaedic Complications

GH treatment moderately increases the risk for slipped femoral capital epiphyses and the risk for worsening of scoliosis. This increased risk seems to be particularly marked in at-risks groups, such as those with Turner syndrome and Prader-Willi (PWS) syndrome and justifies patient information and a diagnosis-orientated clinical surveillance programme.

Otitis Media

GH treatment is associated with a twofold increase of the incidence of otitis media in Turner syndrome in two studies [6, 7], but not in a third one [8]. Given the severity of otological manifestations and their impact on quality of life, all patients with Turner syndrome should be meticulously reviewed by an ENT specialist, particularly when treated with GH.

Sudden Death and Prader-Willi Syndrome

Cases of sudden death have been reported in 28 patients treated with GH [9, 10]. Although the direct implication of GH has not been demonstrated, it is striking that the majority of deaths have occurred within 6 months of starting treatment [10]. Other risks factors, such as morbid obesity, obstructive sleep apnoea and respiratory tract infections, have been associated with these deaths. Altogether, the risk-benefit balance of GH in PWS is highly controversial and it is recommended to perform a sleep apnoea study evaluation and to extensively discuss the risk-benefit balance before considering treatment [11–13].

GH Sensitivity and Prader-Willi Syndrome

Children with PWS appear to show increased sensitivity to GH compared with other GH-treated children. Festen et al. [14] found that total IGF-1 levels often

Table 3. Safety recommendations of GH (somatropin) in the wording of the manufacturers' summary of product characteristics (points taken from all manufacturers' websites and do not relate to any one product)

Contraindications
– Somatropin should not be used in children with closed epiphyses.
– Patients with evidence of progression of an underlying intracranial lesion or other active neoplasms should not receive
   somatropin, since the possibility of a tumour growth-promoting effect cannot be excluded. Prior to the initiation of
   therapy with somatropin, neoplasms must be inactive and anti-tumour therapy completed.
– Hypersensitivity to somatropin or to any of the excipients.
– Patients with acute critical illness suffering complications following open heart surgery, abdominal surgery, multiple
   accidental trauma, acute respiratory failure, or similar conditions should not be treated with somatropin.

Special warnings and precautions for use
– Because somatropin may reduce insulin sensitivity, patients should be monitored for evidence of glucose intolerance.
   For patients with diabetes mellitus, the insulin dose may require adjustment after somatropin therapy is instituted.
   Patients with diabetes or glucose intolerance should be monitored closely during somatropin therapy. Impaired
   glucose tolerance and diabetes mellitus may be unmasked.
– Periodically monitor glucose levels in all patients. Doses of concurrent antihyperglycaemic drugs used in diabetes may
   require adjustment.
– In patients with GH deficiency secondary to an intracranial lesion, frequent monitoring for progression or recurrence
   of the underlying disease process is advised. Discontinue somatropin therapy if progression or recurrence of the
   lesion occurs.
– Patients with Turner's syndrome have an increased risk of developing primary hypothyroidism associated with anti-
   thyroid antibodies.
– There have been reports of sleep apnoea and sudden death associated with the use of GH in paediatric patients with
   PWS who had one or more of the following risk factors: severe obesity, history of respiratory impairment or
   unidentified respiratory infection.
– Scoliosis may progress in any child during rapid growth. Signs of scoliosis should be monitored during somatropin
   treatment.
– Patients with GH failure secondary to chronic kidney disease should be examined periodically for evidence of
   progression of renal osteodystrophy. Slipped capital femoral epiphyses and aseptic necrosis of the femoral head may
   be seen in children with advanced renal osteodystrophy and in GH deficiency, and it is uncertain whether these
   problems are affected by GH therapy.
– Treatment with somatropin should be discontinued at renal transplantation.
– In short children born SGA other medical reasons or treatments that could explain growth disturbance should be
   ruled out before starting treatment.
– In SGA children it is recommended to measure fasting insulin and blood glucose before start of treatment and
   annually thereafter. In patients with increased risk for diabetes mellitus (e.g. familial history of diabetes, obesity, severe
   insulin resistance, acanthosis nigricans) oral glucose tolerance testing should be performed. If overt diabetes occurs,
   GH should not be administered.
– In SGA children it is recommended to measure the IGF-1 level before start of treatment and twice a year thereafter. If
   on repeated measurements IGF-1 levels exceed +2 SD compared to references for age and pubertal status, the IGF-1/
   IGFBP-3 ratio could be taken into account to consider dose adjustment.
– Experience in initiating treatment in SGA patients near onset of puberty is limited. It is therefore not recommended to
   initiate treatment near onset of puberty.
– Experience with patients with Silver-Russell syndrome is limited.

Some of the height gain obtained with treating short children born SGA with GH may be lost if treatment is stopped before final height is reached.

Rare cases of benign intracranial hypertension have been reported. In the event of severe or recurring headache, visual problems, and nausea/vomiting, a funduscopy for papilloedema is recommended. If papilloedema is confirmed, diagnosis of benign intracranial hypertension should be considered and if appropriate GH treatment should be discontinued.

During treatment with somatropin an enhanced T_4 to T_3 conversion has been found which may result in a reduction in serum T_4 and an increase in serum T_3 concentrations. It is therefore particularly advisable to test thyroid function after starting treatment with somatropin and after dose adjustments.

Leukaemia has been reported in a small number of GH-deficient patients treated with somatropin as well as in untreated patients. Based on clinical experience of more than 10 years, the incidence of leukaemia in GH-treated patients without risk factors is not greater than that in the general population.

Slipped capital femoral epiphysis may occur more frequently in patients with endocrine disorders. A patient treated with somatropin who develops a limp or complains of hip or knee pain should be evaluated by a physician.

Concomitant treatment with glucocorticoids inhibits the growth-promoting effects of somatropin. Patients with ACTH deficiency should have their glucocorticoid replacement therapy carefully adjusted to avoid any inhibitory effect on growth.

The effects of treatment with GH on recovery were studied in two placebo-controlled trials involving 522 critically ill adult patients suffering complications following open heart surgery, abdominal surgery, multiple accidental trauma, or acute respiratory failure. Mortality was higher (42 vs. 19%) among patients treated with GHs (doses 5.3–8 mg/day) compared to those receiving placebo. Based on this information, such patients should not be treated with GH. As there is no information available on the safety of GH substitution therapy in acutely critically ill patients, the benefits of continued treatment in this situation should be weighed against the potential risks involved.

Patients who have had GH deficiency as a child should be retested to confirm GH deficiency in adulthood before replacement therapy with somatropin is started.

Undesirable effects

The subcutaneous administration of somatropin may lead to loss or increase of adipose tissue at the injection site. On rare occasions, patients have developed pain and an itchy rash at the site of injection.

Somatropin has given rise to the formation of antibodies in approximately 1% of the patients. The binding capacity of these antibodies has been low and no clinical changes have been associated with their formation.

rose above +2 SDS whereas IGFBP-3 levels remained normal. Sixty to 90% of infants and children had an IGF-1 level >+2 SDS. IGFBP-3 levels remained within the normal range during GH treatment. In this study, GH was well tolerated. Compared with a randomized control group, children with GH treatment did not show disadvantageous effects on carbohydrate metabolism, sleep-related breathing disorders and thyroid hormone levels. Some IGF-1 levels were in the supraphysiological range but were in line with those reported by others in prepubertal PWS children and infants. As PWS children seem to be highly sensitive to GH, it is suggested that monitoring of serum IGF-1 during treatment is routinely performed, as in other situations where GH is used.

Longer Term Safety Considerations

IGF-1 Response to GH Treatment and Its Place in Safety Monitoring
Pretreatment IGF-1 levels are highly variable and range from markedly decreased in severe GH deficiency to normal in non-deficient groups. It had been hoped that both the baseline value and the IGF-1 increase during treatment would be able to predict the growth response, but there is considerable variability in this on account of the difference in IGF-1 sensitivity between various treatment groups. Greater rises are seen in GH deficiency in comparison with GH-sufficient states such as idiopathic short stature (ISS) and children born small for gestational age (SGA) where GH resistance is more in evidence. During the first year of GH treatment, one study showed that relationships between GH dose and the IGF-1 response were weak, and the GH dose for the idiopathic GH-deficient group did not appear to have any influence on IGF-1 SDS at all [15]. Nevertheless, an overdrive of the GH axis is definitely known to have a prolonged effect with IGF-1 levels having been shown to be significantly higher after 5 years of GH treatment in ISS children [16]. The variability of the magnitude of the IGF-1 response and whether this is significant or not makes the prediction of safety of GH treatment using IGF-1 difficult in both the short and long term.

Free versus Total IGF-1
In the circulation, IGF-1 is mainly bound to six classes of IGF-binding proteins. The GH-dependent binding protein 3 (IGFBP-3) is the principal carrier of IGF-1 in plasma, and is responsible for >90% of IGF-1 binding. Usually <1% of IGF-1 in the plasma is unbound. This is the free and biologically active moiety which can exchange rapidly between tissue compartments. It is unclear whether free IGF-1 is a better biomarker of GH activity than either total IGF-1 or IGFBP-3 or the ratio between the two of these. The magnitude of the IGF-1 response to GH treatment in different conditions is also variable, and its use as a predictor of therapeutic success is controversial. A detailed discussion is outside the remit of this chapter. However, is measuring free IGF-1 superior to total IGF-1 in monitoring GH treatment safety? In the only current study on GH treatment (in SGA children), baseline free IGF- levels showed a weak correlation with pretreatment height SDS, but none with any other parameters including spontaneous GH secretion profiles [17]. Pretreatment free IGF-1 levels were not as low as total IGF-1 and IGFBP-3 levels, but during GH therapy free IGF-1 increased inversely related to the growth response but levels remained largely within the normal range, in contrast to the finding of significantly higher total IGF-1 levels, many reaching >2 SDS. In this study there was no dose-dependent rise of either form of IGF-1 as had been reported previously [18]. After the cessation of GH treatment, total and free IGF-1 levels declined, but remained at higher concentrations in comparison with

the pretreatment period. There is clearly a major rise in free IGF-1 levels in this group of patients but the consequences remain unknown. So it remains to be clarified whether it is necessary to adjust the dose of GH to keep the IGF-1 levels (free or total) within the normal range. It is therefore not yet possible to conclude as to whether free IGF-1 is a better biomarker of the safety of long-term GH treatment than evaluation of total IGF-1 levels.

Safety Parameters in Fixed and Variable GH Schedules and IGF-1-Based Dosing
The safety of variable dosage GH regimens has been explored, the intention being to maximize long-term growth but also paying attention to short-term changes in IGF-1 with a concern for safety. This has been explored mostly in non-GH-deficient patients, but it is these groups arguably where there may be more concerns for safety as therapy is more complex than a simple hormone replacement situation. Recent studies in GH-treated SGA children with short stature have explored variable regimens. Gain in height over a 2-year period is similar in standard dose continuous GH treatment in comparison with short-term double- or triple-dose GH treatment [19]. However, the intensive treatment regimens may elevate IGF-1 levels in excess of +3 SDS. The long-term consequences of this degree of elevation, even for periods of 1 year, are unknown.

Adjustment of the GH treatment regimen according to the individual response may be one way forward. A combination of the initial height increment together with metabolic parameters was explored as a way of obtaining optimal growth within safe margins [20]. It has also been proposed that an alternative method of obtaining maximal growth is by titrating GH doses to target IGF-1 levels. Improved growth in GH-treated ISS children was a clear benefit in the study by Cohen et al. [21] where an additional 0.5 SDS in height could be gained after 2 years by targeting the IGF-1 concentration at +2 SDS compared with 0 SDS (the mean) or in comparison with a conventional weight-based GH regimen. However in certain cases a tenfold increment in total GH dose was needed. The consequences of these extremely high GH levels, even where the IGF-1 is at the upper level of normality, are unknown.

Endocrine Consequences of GH Treatment in SGA Children
GH has become widely accepted for growth promotion in short children born SGA. GH therapy produced a significant and sustained rise in mean IGF-1 levels, which in most children was between +1 and +2 SDS during treatment with a standard GH regimen (1 mg/m^2/day) whereas levels rose above +2 SDS during treatment with double this dose per day [22]. Serum IGFBP-3 levels rose significantly during GH treatment but to a lesser extent than IGF-1.

As GH is an insulin antagonist, concerns have also been raised about glucose metabolism in this group of patients who have a natural predisposition to increased insulin resistance and type 2 diabetes mellitus. In clinical trials [23, 24], GH has generally been shown to induce a marked increase in fasting and glucose-stimulated insulin levels and a significant decrease in insulin sensitivity. There was no correlation found between changes in insulin sensitivity during GH treatment and increases in height or height velocity. Therefore, there is no evidence to support the hypothesis that higher insulin levels consequent on GH treatment promote better growth.

Since insulin levels usually rise during puberty, it is hypothesized that children born SGA are more susceptible to GH-induced changes in insulin sensitivity at that time. As more gain in height is obtained during the prepubertal treatment years, it could be argued that GH treatment should not be started close to or during puberty for fear of causing a greater rise in insulin resistance becomes expressed [23].

The question of whether GH treatment affects adrenal function around the time of adrenarche was explored by Hokken et al. [25]. Untreated SGA children had normal serum DHEAS levels before the age of 9 years. The incidence of premature pubarche was comparable with the normal population. One year of GH treatment had no effect on serum DHEAS levels.

Long-term GH treatment of short SGA children did not exert any influence on the age at onset and the progression of puberty compared with appropriate for gestational age controls, irrespective of whether a standard or high-dose GH treatment regimen was used. The duration of puberty and pubertal height gain were similar in both conventional and high GH dosage groups. GH therapy induced considerably higher fasting and glucose-stimulated insulin levels in the short and long term, regardless of dosage regimen. Six months after discontinuation of GH, fasting and glucose-stimulated insulin levels returned to levels comparable with normal, whereas glucose levels were never abnormal during GH treatment.

GH treatment may actually have a positive benefit other than just growth promotion. The Dutch trials showed that untreated short SGA children had higher systolic blood pressure (SBP) but normal diastolic blood pressure (DBP) compared with control SGA children [25]. During 6 years of continuous GH treatment the mean SBP SD score and the lipid atherogenic index decreased significantly.

Long-Term Safety and Risk Factors for Type 2 Diabetes
The effects of GH on increasing insulin resistance in the short term, especially in non-GH-deficient subjects, are well described. However, recent reports in young adults with SGA previously treated with GH show that after ≥6 years after discontinuation of GH, the GH-treated subjects had neither increased fasting glucose levels nor developed type 2 diabetes. Young GH-treated SGA adults had a normal SBP and DBP after stopping GH treatment. In contrast, both SBP and DBP SDS were significantly higher than zero in untreated SGA controls.

These data are reassuring because they suggest that long-term GH treatment does not increase the risk for type 2 diabetes and the metabolic syndrome in young adults [26].

Juvenile Idiopathic Arthritis and Steroids
GH treatment has been shown to be of some benefit for growth promotion and appears to be safe in children with juvenile idiopathic arthritis even with concomitant glucocorticosteroid therapy. A recent randomized controlled trial of high-dose GH treatment started within 2 years of disease onset in children with predominantly systemic-onset juvenile idiopathic arthritis was able to demonstrate the maintenance of growth within the normal range, preventing the usual loss of height seen when GH therapy is not given [27]. The existence of a link between the degree of inflammation and growth impairment was demonstrated by a negative correlation between baseline IGF-1 and CRP, and mean height gain and CRP during follow-up. No significant differences were found in inflammatory markers, or prednisone dosage between the GH and untreated groups. Thus, no worsening of the disease process was evident with treatment and the better growth in the intervention group was ascribed to GH therapy.

Fasting glucose concentrations were not significantly different between the treatment group and the controls over the 3-year follow-up period. Although higher fasting insulin levels were associated with GH treatment, mean HbA1c levels remained within the normal range excepting those with a prednisone dosage >0.7 mg/kg/day which was associated with asymptomatic type 2 diabetes in only 1 patient.

Chronic Kidney Disease
Although GH has been proven to be safe and effective for treatment of uraemic growth failure in later childhood, its usage has not been adequately investigated in infants. Mencarelli et al. [28] report a small study of 7 infants with a height below –2 SDS treated with GH before their first birthday despite adequate caloric intake through a nasogastric tube. Treatment was started in these patients in an attempt to maximize growth so that they would rapidly achieve a suitable size for combined kidney and liver transplantation. The average gain in height SDS was +1.1 ± 0.8 in the treated group and +0.6 ± 1.4 in the control group (p = n.s.) over 2.5 years. Children receiving GH treatment also gained more weight. Assuming that GH improves height by 1 SDS in 3 years, they reported that children would reach a suitable height for transplantation approximately 5 months earlier than they would have done without treatment. Non-fasting serum glucose levels were measured at every outpatient visit and were similar before and after GH treatment.

Parathyroid hormone tended to increase after the initiation of GH therapy. The dose of $1,25(OH)_2$ vitamin D was increased to compensate for this during the first 6 months of GH but after 1 year of treatment, the dose of vitamin D was not found to be different from the pretreatment period. GH therapy may therefore reduce the risk of vascular calcification in both uraemic and transplanted children related to high calcium and phosphate production, high parathyroid hormone levels, and elevated $1,25(OH)_2$ vitamin D dosages in infants with chronic kidney disease.

Effects of GH on the Reproductive Axis
Although no specific concerns about the effect of GH on the reproductive axis have been reported, there are few studies that specifically investigate that. A small but detailed study by Radicioni et al. [29] looked at the long-term effects of GH in non-GH-deficient boys. Investigations including gonadotropins and inhibin B were conducted before and after treatment, the latter phase including semen analysis. In this small study, 7 out of 8 males had a semen analysis within normal limits. Although GH treatment does not generally give rise to concern about the safety on the reproductive axis, this study illustrates one of the few attempts to evaluate an adverse risk.

Tumour Risk and Growth Hormone Treatment

Tumour Recurrence
The use of a mitogenic hormone in the context of cancer raises the question of the risk of tumour recurrence. In the absence of any prospective studies, it is difficult to address the question in a definitive manner. Longitudinal follow-up of large cohorts of GH-treated patients have provided extremely reassuring information on tumour recurrence risk in situations such as neurofibromatosis type 1 [30–32]. Nevertheless, some of the studies have observed that GH-treated patients have a lower risk of recurrence than untreated ones, even after adjustment on risk variables, demonstrating that patients selected for GH treatment have a lower risk of recurrence than untreated ones, a serious limitation on the conclusions of such observational studies [30]. It remains common sense and current practice to refrain from GH treatment in most cases of progressive tumours. After craniopharyngioma surgery for instance, most multidisciplinary teams observe the evolution for at least 6 months before considering GH treatment.

Risk of Second Cancer
In children treated for late effects of cancer, there is an overall substantial risk of a second cancer, in particular meningiomas following cranial irradiation. Here too, evaluating the role of GH in the occurrence of second cancer poses serious methodological difficulties, in the absence of controlled studies [31, 33]. In the long-term observational 'Childhood Cancer Survivor Study', 14,108 children, of whom 361 have been treated with GH, have been followed after a diagnosis of cancer between 1970 and 1986. The risk of second cancer was increased in GH-treated children, after adjustment on covariates (relative incidence 2.15, 95% CI 1.3–3.5; p < 0.002). Of the 361 children treated with GH, 20 developed a secondary solid tumour, among which 9 were meningiomas. The relative risk seems to decrease with time, as there is an overall increase in second cancers in the non-GH-treated

group. Here too, given the design of the study, biases are possible that might explain these observations, in particular the increased surveillance in those who have received GH treatment. It should be remembered that the GH deficiency following cancer treatment is often severe and that replacement generally results in marked benefits including improved quality of life and body composition. It is therefore important to consider all these factors in the overall risk-benefit analysis as to whether GH treatment is going to be of benefit in this situation.

De novo Cancer Risk

In contrast to adverse events occurring during treatment that have been largely documented by several observational studies, there is virtually no data on long-term safety of GH treatment. However, there is a cluster of observations raising the question of de novo cancer risk after GH treatment that emphasizes the need for further studies. In 2002 a cohort study in the UK [34] examined long-term morbidity and mortality in 1,848 patients who had been treated with pituitary-derived GH in childhood. An increased incidence and mortality of colorectal cancer (standardized mortality rate 14.9, 95% CI 1.8–53.9) and of Hodgkin disease (standardized mortality rate 15.3, 95% CI 1.9–55.2) were observed. It should be noted that very few adverse events were observed, explaining the width of the confidence intervals.

Furthermore, there is ample literature on the role of the GH-IGF-1 axis in tumorigenesis [35]. In acromegaly, a situation where GH levels are raised over a prolonged period, the risk of colorectal polyps and cancer is increased. In a meta-analysis, the relative risk was 2.1 (95% CI 1.3–3.1) [36]. In a study performed in Sweden and Denmark, the relative incidence (SIR) was 1.5 for cancers in general (95% CI 1.3–1.8), 2.1 for gastrointestinal cancers (95% CI 1.6–2.7), and specifically 6.0 for small bowel (95% CI 1.2–17.4), 2.6 for colon (95% CI 1.6–3.8), and 2.5 for rectum (95% CI 1.3–4.2). Risks were also increased for tumours of brain (SIR 2.7, 95% CI 1.2–5.0), thyroid (SIR 3.7, 95% CI 1.8–10.9), kidneys (SIR 3.2, 95% CI 1.6–5.5) and bone (SIR 13.8, 95% CI 1.7–50.0). Another study has shown that in acromegalic patients, the risk of colorectal tumour was linked to GH and IGF-1 levels [37]. Similarly, the risk of colorectal polyp recurrence is increased in acromegalic patients [38, 39] and related to IGF-1 levels in some studies [38].

From another perspective, there is an abundance of literature on the influence of IGF-1 levels on the cancer risk in the general population. A meta-analysis has shown an increased risk of colorectal cancer in individuals with the highest IGF-1 levels [40]. The relative risk of colorectal cancer was 1.58 (95% CI 1.11–2.27) for the comparison between those with the highest vs. lowest IGF-1 category. The protective effect of IGFBP-3, which is supposed to 'buffer' the excess IGF-1 in plasma, was not confirmed in this meta-analysis [40].

The Need for New Data: the SAGhE Study
The SAGhE study (Safety and Appropriateness of Growth hormone treatments in Europe) has been designed to address the long-term questions on tolerance that remain currently unanswered [41]. It is a multinational (Belgium, France, Germany, Italy, Netherlands, Sweden, Switzerland, UK) consortium of paediatric endocrinologists and epidemiologists that will prospectively evaluate long-term mortality and morbidity in a large group (≈30,000) of patients who have been treated with GH in childhood (http://saghe.aphp.fr/site/spip.php/). The study has already started in France, based on the national registry of GH-treated patients (Association France-Hypophyse).

Conclusion

Evidence of the long-term safety of GH treatment remains very limited but some of the alarm signals that were initially raised have been shown to be untrue. This should not, however, deter patients, families, physicians and healthcare providers from using a treatment which is effective in the majority of cases. Nevertheless, there is a need for a thorough discussion with patients and families before the start of treatment and they should be asked about conditions resulting in a high risk of cancer such as familial adenomatous polyposis or Li-Fraumeni syndrome. Similarly, decisions about treatment of short stature in conditions such as Fanconi anaemia, Bloom syndrome, neurofibromatosis, where the risk of cancer is increased, should be taken very carefully.

References

1 Consensus: Critical evaluation of the safety of recombinant human growth hormone administration: statement from the Growth Hormone Research Society. J Clin Endocrinol Metab 2001; 86:1868–1870.

2 Cutfield WS, Wilton P, Bennmarker H, Albertsson-Wikland K, Chatelain P, Ranke MB, et al: Incidence of diabetes mellitus and impaired glucose tolerance in children and adolescents receiving growth-hormone treatment. Lancet 2000;355: 610–613.

3 Libman IM, Barinas-Mitchell E, Bartucci A, Robertson R, Arslanian S: Reproducibility of the oral glucose tolerance test in overweight children. J Clin Endocrinol Metab 2008;93:4231–4237.

4 Roman R, Zeitler PS: Oral glucose tolerance testing in asymptomatic obese children: more questions than answers. J Clin Endocrinol Metab 2008;93:4228–4230.

5 Malozowski S, Tanner LA, Wysowski DK, Fleming GA, Stadel BV: Benign intracranial hypertension in children with growth hormone deficiency treated with growth hormone. J Pediatr 1995;126: 996–999.

6 Quigley CA, Crowe BJ, Anglin DG, Chipman JJ: Growth hormone and low dose estrogen in Turner syndrome: results of a United States multi-center trial to near-final height. J Clin Endocrinol Metab 2002;87:2033–2041.

   Carel · Butler

7   Stephure DK: Impact of growth hormone supplementation on adult height in Turner syndrome: results of the Canadian randomized controlled trial. J Clin Endocrinol Metab 2005;90:3360–3366.

8   Davenport ML, Crowe BJ, Travers SH, Rubin K, Ross JL, Fechner PY, et al: Growth hormone treatment of early growth failure in toddlers with Turner syndrome: a randomized, controlled, multicenter trial. J Clin Endocrinol Metab 2007;92:3406–3416.

9   Eiholzer U: Deaths in children with Prader-Willi syndrome. A contribution to the debate about the safety of growth hormone treatment in children with PWS. Horm Res 2005;63:33–39.

10 Tauber M, Diene G, Molinas C, Hebert M: Review of 64 cases of death in children with Prader-Willi syndrome. Am J Med Genet A 2008;146:881–887.

11 Fillion M, Deal C, Van Vliet G: Retrospective study of the potential benefits and adverse events during growth hormone treatment in children with Prader-Willi syndrome. J Pediatr 2009;154:230–233.

12 Tauber M, Hokken-Koelega AC, Hauffa BP, Goldstone AP: About the benefits of growth hormone treatment in children with Prader-Willi syndrome. J Pediatr 2009;154:778–779.

13 Stafler P, Wallis C: Prader-Willi syndrome: who can have growth hormone? Arch Dis Child 2008;93:341–345.

14 Festen DA, de Lind van Wijngaarden R, van Eekelen M, Otten BJ, Wit JM, Duivenvoorden HJ, Hokken-Koelega AC: Randomized controlled GH trial: effects on anthropometry, body composition and body proportions in a large group of children with Prader-Willi syndrome. Clin Endocrinol 2008;69:443–451.

15 Cutfield W, Lundgren F: Insulin-like growth factor 1 and growth responses during the first year of growth hormone treatment in KIGS patients with idiopathic growth hormone deficiency, acquired growth hormone deficiency, turner syndrome and born small for gestational age. Horm Res 2009;71(suppl 1):39–45.

16 Quigley CA, Gill AM, Crowe BJ, Robling K, Chipman JJ, Rose SR, Ross JL, Cassorla FG, Wolka AM, Wit JM, Rekers-Mombarg LT, Cutler GB: Safety of growth hormone treatment in pediatric patients with idiopathic short stature. J Clin Endocrinol Metab 2005;90:5188–5196.

17 Bannink EM, van Doorn J, Mulder PG, Hokken-Koelega AC: Free/dissociable insulin-like growth factor (IGF)-1, not total IGF-1, correlates with growth response during growth hormone treatment in children born small for gestational age. J Clin Endocrinol Metab 2007;92:2992–3000.

18 Ranke MB, Traunecker R, Martin DD, Schweizer R, Schwarze CP, Wollmann HA, Binder G: IGF-1 and IGF binding protein-3 levels during initial GH dosage step-up are indicators of GH sensitivity in GH-deficient children and short children born small for gestational age. Horm Res 2005;64:68–76.

19 Phillip M, Lebenthal Y, Lebl J, Zuckerman-Levin N, Korpal-Szczyrska M, Marques JS, Steensberg A, Jons K, Kappelgaard AM, Ibanez L, European Norditropin SGASG: European multicentre study in children born small for gestational age with persistent short stature: comparison of continuous and discontinuous growth hormone treatment regimens. Horm Res 2009;71:52–59.

20 Jung H, Land C, Nicolay C, De Schepper J, Blum WF, Schonau E: Growth response to an individualised versus fixed dose GH treatment in short children born small for gestational age: the OPTIMA study. Eur J Endocrinol 2009;160:149–156.

21 Cohen P, Rogol AD, Howard CP, Bright GM, Kappelgaard AM, Rosenfeld RG: Insulin growth factor-based dosing of growth hormone therapy in children: a randomized, controlled study. J Clin Endocrinol Metab 2007;92:2480–2486.

22 Hokken-Koelega AC, van Pareren YK, Arends N, Boonstra V: Efficacy and safety of long-term continuous growth hormone treatment of children born small for gestational age. Horm Res 2004;62(suppl 3):149–154.

23 Bachmann S, Bechtold S, Bonfig W, Putzker S, Buckl M, Schwarz HP: Insulin sensitivity decreases in short children born small for gestational age treated with growth hormone. J Pediatr 2009;154:509–513.

24 Cutfield WS, Regan FA, Jackson WE, Jefferies CA, Robinson EM, Harris M, Hofman PL: The endocrine consequences for very low birth weight premature infants. Growth Horm IGF Res 2004;14:S130–S135.

25 Hokken-Koelega A, van Pareren Y, Arends N, Boonstra V: Efficacy and safety of long-term continuous growth hormone treatment of children born small for gestational age. Horm Res 2004;62(suppl 3):149–154.

26 Van Dijk M, Bannink EM, van Pareren YK, Mulder PG, Hokken-Koelega AC: Risk factors for diabetes mellitus type 2 and metabolic syndrome are comparable for previously growth hormone-treated young adults born small for gestational age (SGA) and untreated short SGA controls. J Clin Endocrinol Metab 2007;92:160–165.

27 Simon D, Prieur AM, Quartier P, Charles Ruiz J, Czernichow P: Early recombinant Human growth hormone treatment in glucocorticoid-treated children with juvenile idiopathic arthritis: a 3-year randomized study. J Clin Endocrinol Metab 2007; 92:2567–2573.

28 Mencarelli F, Kiepe D, Leozappa G, Stringini G, Cappa M, Emma F: Growth hormone treatment started in the first year of life in infants with chronic renal failure. Pediatr Nephrol 2009;24: 1039–1046.

29 Radicioni AF, Paris, E, De Marco E, Anzuini A, Gandini L, Lenzi A: Testicular function in boys previously treated with recombinant human growth hormone for non-growth hormone-deficient short stature. J Endocrinol Invest 2007;30: 931–936.

30 Swerdlow AJ, Reddingius RE, Higgins CD, Spoudeas HA, Phipps K, Qiao Z, et al: Growth hormone treatment of children with brain tumors and risk of tumor recurrence. J Clin Endocrinol Metab 2000;85:4444–4449.

31 Sklar CA, Mertens AC, Mitby P, Occhiogrosso G, Qin J, Heller G, et al: Risk of disease recurrence and second neoplasms in survivors of childhood cancer treated with growth hormone: a report from the Childhood Cancer Survivor Study. J Clin Endocrinol Metab 2002;87:3136–3141.

32 Jenkins PJ, Mukherjee A, Shalet SM: Does growth hormone cause cancer? Clin Endocrinol (Oxf) 2006;64:115–121.

33 Ergun-Longmire B, Mertens AC, Mitby P, Qin J, Heller G, Shi W, et al: Growth hormone treatment and risk of second neoplasms in the childhood cancer survivor. J Clin Endocrinol Metab 2006; 91:3494–3498.

34 Swerdlow AJ, Higgins CD, Adlard P, Preece MA: Risk of cancer in patients treated with human pituitary growth hormone in the UK, 1959–1985: a cohort study. Lancet 2002;360:273–277.

35 Pollak MN, Schernhammer ES, Hankinson SE: Insulin-like growth factors and neoplasia. Nat Rev Cancer 2004;4:505–518.

36 Renehan AG, O'Connell J, O'Halloran D, Shanahan F, Potten CS, O'Dwyer ST, et al: Acromegaly and colorectal cancer: a comprehensive review of epidemiology, biological mechanisms, and clinical implications. Horm Metab Res 2003;35:712–725.

37 Orme SM, McNally RJ, Cartwright RA, Belchetz PE: Mortality and cancer incidence in acromegaly: a retrospective cohort study. United Kingdom Acromegaly Study Group. J Clin Endocrinol Metab 1998;83: 2730–2734.

38 Jenkins PJ, Frajese V, Jones AM, Camacho-Hubner C, Lowe DG, Fairclough PD, et al: Insulin-like growth factor 1 and the development of colorectal neoplasia in acromegaly. J Clin Endocrinol Metab 2000;85:3218–3221.

39 Terzolo M, Reimondo G, Gasperi M, Cozzi R, Pivonello R, Vitale G, et al: Colonoscopic screening and follow-up in patients with acromegaly: a multicenter study in Italy. J Clin Endocrinol Metab 2005;90:84–90.

40 Renehan AG, Zwahlen M, Minder C, O'Dwyer ST, Shalet SM, Egger M: Insulin-like growth factor (IGF)-1, IGF-binding protein-3, and cancer risk: systematic review and meta-regression analysis. Lancet 2004;363:1346–1353.

41 Safety and Appropriateness of Growth hormone treatments in Europe (SAGhE): http://saghe.aphp. fr/site/spip.php/

Gary Butler
Department of Paediatrics and Adolescents
University College London Hospital
250 Euston Road, London NW1 2PQ (UK)
Tel. +44 8451 555 000, E-Mail g.butler@ich.ucl.ac.uk

Hindmarsh PC (ed): Current Indications for Growth Hormone Therapy, ed 2, revised.
Endocr Dev. Basel, Karger, 2010, vol 18, pp 55–66

Diagnosis of Growth Hormone Deficiency

E.A. Webb · M.T. Dattani

Developmental Endocrinology Research Group, UCL Institute of Child Health, and Department of
Endocrinology, Great Ormond Street Children's Hospital, London, UK

Abstract

The diagnosis of growth hormone deficiency (GHD) was essentially a clinical one prior to the
advent of radioimmunoassay in the mid-1960s. From this point on both clinical and biochemical serum GH responses to a variety of provocation tests were used to define the condition. The
definition of an adequate GH response to stimulation has changed over time, initially being <3
µg/l and gradually increasing to 10 µg/l. Over this period, GH assays became more sensitive and
specific, and assays for IGF-1 and IGFBP-3 were developed. Detailed neuroimaging also became
widely available and genetic aetiologies for GHD identified. Apart from a number of clear
genetic causes for GHD, the diagnostic gold standard remains elusive. However, making the
correct diagnosis has significant benefits to the child, guiding both future investigations and
management. In this chapter we discuss the importance of taking into account all available
evidence when making the diagnosis of GHD including clinical examination, detailed auxological measurements, bone age, molecular analysis, biochemical measures and neuroradiological
assessment.

Introduction

Growth hormone (GH) is an anabolic protein secreted by the somatotroph cells of
the anterior pituitary gland in a pulsatile fashion [1]. The main isoform of human
GH is a 191-amino-acid protein (molecular weight 22 kDa) which acts to stimulate
linear growth, increase muscle strength and bone density. Both excess secretion
and deficiency of the hormone produce disease states, acromegaly and gigantism
in the case of the former, and poor growth and short stature with respect to the
latter. The reported incidence of congenital GH deficiency (GHD) is 1 in 4,000 to
1 in 10,000 live births. GHD occurs more frequently in males, and can be familial,
idiopathic or acquired [2, 3] (table 1).

GH was first identified in rats in 1922 [4], however it was not until 1958 that
Maurice Raben [5] successfully purified enough cadaveric GH to treat GHD.

Table 1. Aetiology of GHD

Congenital	
Genetic	Isolated pituitary deficiency, multiple pituitary hormone deficiencies, syndromes associated with pituitary hormone abnormalities (see table 2)
Acquired	
Trauma	Perinatal, post-natal
Infection	Meningitis, encephalitis
Langerhans cell histocytosis	
CNS tumours	Compression effect Post-cranial irradiation
After chemotherapy	
Psychosocial deprivation	
Neurosecretory dysfunction	
Hypothyroidism	

Initially the diagnosis was a clinical one until the advent of radioimmunoassay in the mid-1960s. From this point on, both clinical and biochemical serum GH responses to a variety of provocation tests were used to define the condition. As supplies of cadaveric GH were limited, only patients with severe GHD (peak GH <3 µg/l) were eligible for replacement therapy. When recombinant GH became available in 1985, such a restrictive approach was less necessary. The cut-off point for GH treatment was increased arbitrarily, without the availability of normative data, initially to 7 µg/l and subsequently to 10 µg/l [6, 7]. Over this period, GH assays became more sensitive and specific, and assays for insulin-like growth factor-1 (IGF-1) and its binding protein (IGFBP-3) were developed. Detailed neuroimaging also became widely available and genetic aetiologies for GHD identified [8–10]. Apart from a number of clear genetic causes for GHD, the diagnostic gold standard remains elusive [11].

GH stimulation tests remain widely used to diagnose GHD yet they can be dangerous, difficult to interpret, and lack sensitivity and specificity [12, 13]. It remains to be defined what constitutes a normal GH response to stimulation testing and therefore the criteria that should be used to identify those individuals with GHD requiring GH replacement [14]. Currently the criteria used for diagnosis vary from centre to centre, with some centres no longer performing provocative testing, arguing that both GHD and non-GHD children with low IGF-1 or structural hypothalamo-pituitary abnormalities respond to treatment with GH [15].

However, making the correct diagnosis has significant benefits to the child, guiding both future investigations and management. Children with GHD frequently present with short stature but can go on to develop other potentially life-threatening pituitary hormone abnormalities, and are at higher risk of having other central nervous system (CNS) abnormalities [16, 17]. Individuals with non-GHD short stature may have another underlying diagnosis causing their poor growth and therefore warrant more rigorous investigation of other symptoms [13].

In this chapter we discuss the importance of taking into account all available evidence when making the diagnosis of GHD, including clinical examination, detailed auxological measurements, bone age, molecular analysis, biochemical measures and neuroradiological assessment.

Clinical Examination and Auxology

Clinical examination and rigorous auxological measurements remain vital both to making the initial decision regarding which individuals require investigation for GHD, and to making the diagnosis of GHD. A detailed clinical examination may reveal typical phenotypic features of GHD including: frontal bossing, mid-facial hypoplasia, immature facies, truncal adiposity, fat dimpling, male hypogenitalism, a high-pitched voice and/or other malformations. Taking a comprehensive history may also identify children at increased risk of having GHD with peri- and post-natal trauma, CNS tumours, CNS irradiation, chemotherapy, history of consanguinity and/or other siblings with GHD all being associated with increased risk of GH insufficiency.

Where other associated malformations are present or where there is a family history of significant short stature, an identifiable genetic aetiology is more likely to be found (table 1) [18]. Although currently genetic abnormalities are only found in approximately 10% of individuals with GHD, the identification of a genetic mutation within the GH-1 or GHRHR genes, or one of the transcription factors associated with multiple pituitary hormone deficiency, can prove the diagnosis (table 2) [19]. First-degree relatives are affected in 5–30% of children with GHD, and 40–60% of individuals with sporadic GHD have an abnormal hypothalamo-pituitary axis on neuroimaging, suggesting that there may be further, as yet unidentified, genetic aetiologies for GHD [20, 21].

There may also be clues regarding the diagnosis in the medical history, with neonatal hypoglycaemia, jaundice and micropenis all being more common in those with GHD [21]. However, the most sensitive marker of GHD on clinical examination remains growth velocity [22]. The main clinical characteristic of GHD is the child presenting with a reduced height velocity, delayed bone age and ultimately a delayed entrance into puberty. In children with congenital GHD, growth failure is present

Table 2. Human gene abnormalities causing abnormal hypothalamo-pituitary development and function

Gene	Phenotype	Inheritance
Isolated growth hormone deficiency		
GH-1	IGHD. No response to GH stimulation	Recessive, dominant
GHRHR	IGHD	Recessive
Combined pituitary hormone deficiency (CPHD)		
POU1F1	GH, TSH, prolactin deficiencies; usually severe; small or normal AP	Recessive, dominant
PROP1	GH, TSH, LH, FSH, prolactin deficiencies; evolving ACTH deficiency; small, normal or enlarged AP	Recessive
Specific syndrome		
HESX1	IGHD, CPHD, septo-optic dysplasia; APH, EPP, absent infundibulum, ONH, ACC	Recessive, dominant
LHX3	CPHD (GH, TSH, LH, FSH, prolactin and ACTH deficiencies), short neck, limited rotation; small, normal or enlarged AP, short cervical spine, sensorineural hearing loss	Recessive
LHX4	CPHD (GH, TSH, ACTH deficiencies); small AP, EPP, cerebellar abnormalities	Dominant
SOX3	IGHD and mental retardation, panhypopituitarism; APH, infundibular hypoplasia, EPP	X Linked
SOX2	Hypogonadotrophic hypogonadism; APH, bilateral anophthalmia/microphthalmia, abnormal corpus callosum, hypothalamic hamartoma, learning difficulties, oesophageal atresia, sensorineural hearing loss	De novo
GLI2	Holoprosencephaly and multiple midline defects	AD
OTX2	CPHD anophthalmia, small anterior and ectopic posterior pituitary	AD

AP(H) = Anterior pituitary (hypoplasia); EPP = ectopic posterior pituitary; ACC = agenesis of corpus callosum, ONH = optic nerve hypoplasia.

from early in life [23]. At any age, growth failure with an inappropriate height velocity for age (<3rd centile height velocity calculated over an interval of not less than 6 months), with a height centile not predicted by mid-parental target height warrants further investigation. In this scenario the probability of a subsequent normal

12-month height velocity is 3% [13]. Calculating the weight-to-height ratio is also helpful, as children with GHD tend to have a relatively preserved weight. On the other hand, children with other aetiologies for their poor growth such as coeliac disease, inflammatory bowel disease, chronic renal or liver disease and cystic fibrosis tend to have poor weight gain in parallel with their poor linear growth.

Measuring GH Secretion

When GH was first administered to patients in 1958, no assay for GH existed, and the diagnosis of GHD was made using auxological criteria, in conjunction with clinical examination and biochemical measurements (e.g. urinary hydroxyproline). During the 1960s the advent of radioimmunoassay allowed for the measurement of GH in the circulation. Two main forms of GH were subsequently identified, 22 and 20 kDa GH. These isoforms account for most of the GH released in response to GH-releasing hormone stimulation (70–75 and 5–10% of total secretion respectively) [24–26]. Numerous other forms have also been measured but despite this there has been a move towards using assays that only measure 22 kDa GH [27].

In parallel with the immunoassay development, clinical and physiological studies demonstrated that GH was predominantly secreted overnight in a pulsatile manner, with concentrations normally undetectable during the day. The implication of this is that a single random measurement of GH is an unhelpful diagnostic test, except in neonates, in whom a random GH of <20 µg/l is suggestive of GHD [28]. Quantifying overnight GH release is both time- and labour-intensive, requiring collection of serum samples every 20 min for a minimum of 12 h, and may not identify all those with GHD [29]. This is partly because intra-individual variation is considerable, and also because the lack of normative data pertaining to age, sex, body mass index and pubertal status makes interpretation difficult [30, 31]. In addition, significant overlap exists between spontaneous GH production in individuals with and without GHD. 25% of normally growing children having low overnight GH production [32]. Consequently, a variety of GH stimulation tests, both physiological (including sleep, exercise, fasting and hypoglycaemia) and pharmacological (such as glucagon, insulin, arginine and clonidine) have been developed to determine an individual's capacity for GH release.

Provocative testing remains the most widely used and accepted method of diagnosing GHD. Currently there are at least 34 provocative tests used in children, with 189 different combinations of these tests [20, 33]. There is a huge literature regarding the value of these stimulation tests in children from which it is impossible to draw firm conclusions. The 'cut-off' peak serum GH concentration used to define GHD varies considerably and is dependent on the assay used and stimuli applied to determine peak GH response. Additionally, the criteria used for the diagnosis of certain GHD (e.g. slow growth rate, levels of indirect markers of GH production, MRI abnormalities), vary significantly between studies. Severe/complete GHD has classically been defined as a peak serum GH response to stimulation of <5 μg/l and moderate/partial GHD as <10 μg/l, with no differentiation made to account for the strength of the stimulation test used to reach the diagnosis. Most studies have found those with complete GHD to have a more consistent response to provocative testing and it is therefore the diagnosis of partial GHD which is particularly hard to reach using current protocols [29].

No stimulation test performed in isolation provides adequate specificity for the diagnosis of GHD, and therefore, it is generally advised that at least two provocative tests are performed to improve sensitivity and specificity, although little evidence exists to support these recommendations [29]. Whilst ensuring that the diagnosis is more certain by insisting that two tests must be low for diagnosis it does increase the number of individuals in the grey zone where one test may be diagnostic but the other of borderline significance.

There are inherent problems in performing two tests simultaneously or for that matter separated in time. Consecutive tests following on from each other may be effected by down-regulation of the hypothalamo-pituitary axis, with repeated stimulation reducing and altering the proportions of GH isoforms released [34]. The results of a second test during a 24-hour period should therefore be interpreted with caution [33] and certainly it is not reasonable to apply the same cut-off criteria to the second test as to the first. Performing a second test on a separate day is also problematic as the tests can be clearly shown to be independent of each other and as a result invokes a series of complex statistical manoeuvres to accommodate the findings. Manoeuvres that are seldom undertaken in the clinical situation.

Even with the multitude of tests available, there is little consistent evidence pertaining to their reproducibility and reliability in making the diagnosis of GHD [34], with many normal children failing stimulation tests, especially in the prepubertal period or in those with obesity [35, 36]. GH pulse amplitude increases with puberty and several studies have shown that children who have failed previous GH tests have normal responses after they are primed with sex steroids [37, 38]. However, the role of sex-steroid priming in prepubertal children remains controversial, with the interpretation of these studies still debated [29].

The variation between studies regarding sensitivity and specificity of GH provocations tests also reflects the lack of standardized assays for GH measurement. Fully automated two-site sandwich immunoassays have, in the majority of laboratories, superseded the manual radioimmunoassay, with the former providing significantly lower results than the latter [33]. Surprisingly however, the recommended 'cut-off' for normal GH response to stimulation has increased since the introduction of sandwich immunoassays. The variability in GH results can be as much as 200% between assays and is likely due to the molecular heterogeneity of GH, interference from

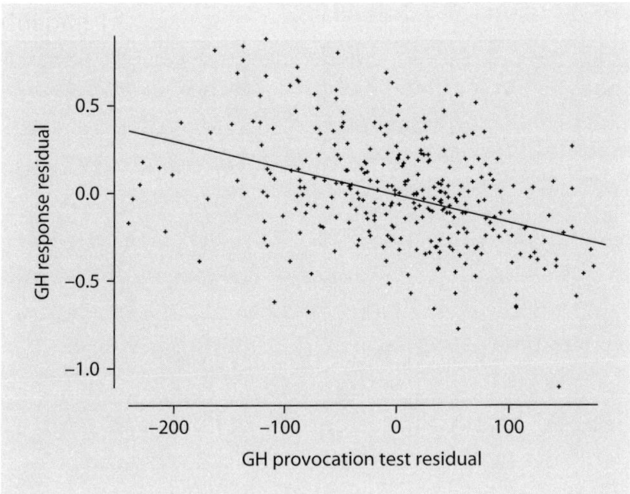

Fig. 1. Relation between height SDS change in the first year of GH treatment and maximum GH peak (%), each expressed as residuals.

matrix components such as GH-binding protein and the differences in GH standards, diluents and matrices used [39]. One study found that when peak samples were re-assayed centrally using one reference assay, out of 132 patients, 35 were re-categorized as having GHD [40]. Reliable analysis of results from GH testing therefore depends on each laboratory performing GH assays collating normative data relating to age, sex, body composition, pubertal stage and test used to stimulate GH, and assigning their own 'normal cut-off' values for the diagnosis of GHD based on this information [41]. In practice however, collecting the amount of normative data required to allow interpretation of results is not feasible, and perhaps a better solution would be for all centres to use a standardized protocol and GH assay [13, 39].

To try and overcome these problems, numerous other methods of measuring GH production have been evaluated. Measurement of urinary GH excretion, overnight GH production, metabolic responses to short-term GH administration and growth response to GH treatment have all been investigated [42]. Unfortunately the deficiencies of GH stimulation tests have not been overcome by using any of these methods, although the peak GH response to provocative testing is predictive of subsequent growth response to GH (fig. 1). Not all patients who appear to be GHD on provocative testing subsequently have growth acceleration when GH treatment is started [34, 43]. Cole et al. [34] therefore recommended basing initial GH treatment on results of provocative testing and if the subsequent height velocity on GH after 12 months is sufficiently low, and compliance with medication is adequate, reviewing the diagnosis at that time.

Using Downstream Targets of GH Action – IGF-1 and IGFBP-3

Since a specific radioimmunoassay was first developed for IGF-1 and subsequently for IGFBP-3, there has been significant interest in their use as markers for GHD. Serum IGF-1 and IGFBP-3 were thought to be ideal markers of GHD as they are indirect markers of GH secretion, can be measured in the fed or fasted state, and do not have a circadian rhythm. Both IGF-1 and IGFBP-3 have been proposed as a means to 'screen' patients with short stature for underlying hormonal abnormalities [44, 45]. However, given the poor diagnostic performance of all tests along with the low prevalence of the condition this is perhaps optimistic. That said, several studies have shown that low concentrations of serum IGF-1, IGF-2 and IGFBP-3 and high concentrations of serum IGFBP-2 are predictive of GHD, but not all studies have confirmed these findings [46, 47].

Although some studies have found a serum IGF-1 concentration <2 SDs below the mean to be highly specific for the diagnosis of GHD, 30% of children with normal IGF-1 measurements also have GHD. The differences found in the sensitivity and specificity of IGF-1 for the diagnosis of GHD is partly due to the significant overlap in serum IGF-1 between short and normal stature children and also due to the variation in IGF-1 concentrations with body mass index [7, 29, 46–52]. It also reflects the lack of standardization between the studies themselves, with different stimuli, assays for GH, IGF-1 and IGFBP-3 and the normative data used to interpret the results of the tests used. Serum IGF-1 concentrations are also low in children with malnutrition, chronic liver or renal disease and hypothyroidism [53]. Concentrations of IGF-1 also vary with age, pubertal status and sex, and therefore need to be standardized for these variables [54, 55].

Serum IGFBP-3 concentrations (standardized for age and sex) are highly specific for the diagnosis of GHD but lack sensitivity, with normal concentrations not ruling out the diagnosis of GHD. Low serum IGFBP-3 concentrations are therefore highly suggestive of GHD, except in the scenarios of GH insensitivity or liver disease. Serum IGF-1 or IGFBP-3 measured in isolation cannot therefore be used to discriminate between patient groups, although diagnostic specificity increases when IGF-1 and IGFBP-3 are measured in conjunction with stimulation testing [17].

Discordance between results from provocative testing and measurements of IGF-1 and IGFBP-3 may suggest diagnoses other than straightforward GHD. For example, individuals with above average GH response to stimulation but values for IGF-1 and IGFBP-3 less than –2 SDs below the normal range may have GH insensitivity and require further specialized investigations to elucidate the diagnosis [56].

Neuroradiology

Magnetic resonance imaging (MRI) is the most sensitive imaging modality available for visualizing the hypothalamo-pituitary axis. It allows delineation of the anatomy of the hypothalamo-pituitary region, any associated developmental abnormalities such as optic nerve hypoplasia or dysgenesis of the corpus callosum and the identification of tumours which may be the primary cause of the growth failure.

Using the presence or absence of any MRI abnormalities to confirm GHD yields low sensitivity and specificity (79% and 54% respectively) [46]. However, the presence of either an undescended or ectopic posterior pituitary gland or a hypoplastic anterior pituitary gland in association with a hypoplastic or absent pituitary stalk is highly specific (100% and 89% respectively) and predictive of GHD (positive predictive value 100% and 79% respectively) [17]. MRI can therefore be used to confirm (or refute) the diagnosis of GHD in some individuals with biochemical, clinical and auxological data suggestive of the diagnosis. It can also be used to guide ongoing management of these children as individuals with a more severe MRI phenotype, (absent infundibulum, undescended/ectopic posterior pituitary positioned at the median eminence as opposed to along the pituitary stalk), are more likely to have persistent GHD associated with other pituitary hormone deficiencies than those with an isolated undescended/ectopic posterior pituitary [17].

Children with certain MRI abnormalities are at lifelong risk of developing other pituitary hormone abnormalities even if GH peak to stimulation testing at final height is normal [57]. This finding suggests that somatotroph function can recover at least temporarily, and that the documentation of a normal GH response to provocation at the end of growth in 26% of individuals may not necessarily indicate failure of the original diagnostic test [58].

Conclusions

The diagnosis of GHD is perhaps one of the most controversial and hotly debated issues in endocrinology. One of the major difficulties of defining GHD is that there is a continuum between deficiency and normal secretion. The situation is rather analogous to the hypertension argument as to when a higher than average becomes pathological. In the case of blood pressure there are good epidemiological data to indicate this point. With GH those data are not available but judgement could be made, albeit subjective to a degree, on the subsequent growth acceleration resulting from the introduction of GH therapy.

The current gold standard for the diagnosis of GHD, as defined by the National Institute for Clinical Excellence (NICE) guidelines, remains a subnormal response to two pharmacological stimulation tests. The diagnosis can be made with a high

42 Ranke MB, Lindberg A, Chatelain P, et al: Derivation and validation of a mathematical model for predicting the response to exogenous recombinant human growth hormone (GH) in prepubertal children with idiopathic GH deficiency. KIGS International Board. Kabi Pharmacia International Growth Study. J Clin Endocrinol Metab 1999;84: 1174–1183.

43 Aimaretti G, Baffoni C, Bellone S, et al: Retesting young adults with childhood-onset growth hormone (GH) deficiency with GH-releasing-hormone-plus-arginine test. J Clin Endocrinol Metab 2000;85:3693–3699.

44 Loche S, Bizzarri C, Maghnie M, et al: Results Of early re-evaluation of growth hormone secretion in short children with apparent growth hormone deficiency. J Pediatr 2002;140:445–449.

45 Mitchell H, Dattani MT, Nanduri V, Hindmarsh PC, Preece MA, Brook CG: Failure of IGF-1 and IGFBP-3 to diagnose growth hormone insufficiency. Arch Dis Child 1999;80:443–447.

46 Blum WF, Ranke MB, Kietzmann K, Gauggel E, Zeisel HJ, Bierich JR: A specific radioimmunoassay for the growth hormone (GH)-dependent somatomedin-binding protein: its use for diagnosis of GH deficiency. J Clin Endocrinol Metab 1990;70:1292–1298.

47 Cianfarani S, Boemi S, Spagnoli A, et al: Is IGF binding protein-3 assessment helpful for the diagnosis of GH deficiency? Clin Endocrinol (Oxf) 1995;43:43–47.

48 Smith WJ, Nam TJ, Underwood LE, Busby WH, Celnicker A, Clemmons DR: Use of insulin-like growth factor-binding protein (IGFBP)-2, IGFBP-3, AND IGF-1 for assessing growth hormone status in short children. J Clin Endocrinol Metab 1993;77:1294–1299.

49 Tillmann V, Buckler JM, Kibirige MS, et al: Biochemical tests in the diagnosis of childhood growth hormone deficiency. J Clin Endocrinol Metab 1997;82:531–535.

50 Sklar C, Sarafoglou K, Whittam E: Efficacy of insulin-like growth factor binding protein 3 in predicting the growth hormone response to provocative testing in children treated with cranial irradiation. Acta Endocrinol (Copenh) 1993;129: 511–515.

51 Phillip M, Chalew SA, Kowarski AA, Stene MA: Plasma IGFBP-3 and its relationship with quantitative growth hormone secretion in short children. Clin Endocrinol (Oxf) 1993;39:427–432.

52 Rasmussen MH, Frystyk J, Andersen T, Breum L, Christiansen JS, Hilsted J: The impact of obesity, fat distribution, and energy restriction on insulin-like growth factor-1 (IGF-1), IGF-binding protein-3, insulin, and growth hormone. Metabolism 1994;43:315–319.

53 Juul A, Skakkebaek NE: Prediction of the outcome of growth hormone provocative testing in short children by measurement of serum levels of insulin-like growth factor I and insulin-like growth factor binding protein 3. J Pediatr 1997; 130:197–204.

54 Juul A: Determination of insulin-like growth factor 1 In children: normal values and clinical use. Horm Res 2001;55(suppl 2):94–99.

55 Laron Z: Growth hormone insensitivity (Laron syndrome). Rev Endocr Metab Disord 2002;3: 347–355.

56 Leger J, Danner S, Simon D, Garel C, Czernichow P: Do all patients with childhood-onset growth hormone deficiency (GHD) and ectopic neurohypophysis have persistent GHD in adulthood? J Clin Endocrinol Metab 2005;90:650–656.

57 Clayton PE, Price DA, Shalet SM: Growth hormone state after completion of treatment with growth hormone. Arch Dis Child 1987;62:222–226.

Mehul Dattani, MD
Developmental Endocrinology Research Group, Clinical and Molecular Genetics Unit
Institute for Child Health, University College London
30 Guilford Street, London WC1N 1EH (UK)
Tel. +44 207 905 2657, Fax +44 207 404 6191, E-Mail e.webb@ich.ucl.ac.uk

Hindmarsh PC (ed): Current Indications for Growth Hormone Therapy, ed 2, revised.
Endocr Dev. Basel, Karger, 2010, vol 18, pp 67–82

Biological Determinants of Responsiveness to Growth Hormone: Pharmacogenomics and Personalized Medicine

Primus-E. Mullis

Paediatric Endocrinology, Diabetology and Metabolism, University Children's Hospital, Inselspital, Bern, Switzerland

Abstract

It is becoming most clear that many genes are involved in controlling the regulation of growth. Ultimately however, at the level of growth hormone (GH), the relevant question may be not whether a patient is GH-deficient, but whether he is GH-responsive. As these disturbances can be divided into two gross categories, namely alterations causing subnormal GH secretion and/or those presenting with subnormal GH sensitivity/responsiveness, the main aim of this review is to focus on genes involved in growth regulation leading to short stature caused by an alteration of GH insensitivity/GH responsiveness; in other words, clinical circumstances where individually adapted GH replacement therapy may help to increase height velocity and eventually final height.   Copyright © 2010 S. Karger AG, Basel

Introduction

The most fundamental characteristic of infancy as well as childhood is growth. Although the process of growth is multifactorial and complex, the growth pattern of children, if evaluated in the context of normal standards, is rather predictable. In Switzerland for instance, the growth charts of the First Zurich Longitudinal Study of Growth and Development, where data are expressed as SDS, are used and serve the process of identifying abnormal growth well [1]. Height in a population follows approximately a gaussian distribution, similar to many other polygenetic traits. Any deviation from a normal pattern of growth can be the first manifestation of a wide variety of disease processes, including endocrine and non-endocrine disorders and, importantly, may involve any organ system of the human body.

For a considerable period of time, growth disorders were managed on the basis of a growth hormone (GH)-oriented classification system. Nowadays, however, clinicians are well aware that (a) GH on its own is not the major mediator of skeletal growth, (b) the tests used to diagnose GH deficiency (GHD) have many problems (see Chapter 1), and (c) many genetic defects have been described and have presented important insights into the molecular basis of GHD and non-GHD growth failure [2]. Furthermore, as pediatric endocrinologists we are regularly confronted with children presenting with short stature who are given the label of idiopathic short stature (ISS). ISS is a purely descriptive term referring to a child with a height below the age reference for population and sex, in whom with our diagnostic tools no etiological diagnosis has been made and in whom it may have to be accepted that this is the extreme of the normal distribution. Unlike hypertension there is no firm end-point to measure against so placing a cut-point for height normality becomes difficult.

One of the best known problems is to distinguish ISS from 'partial' GHD. There is no 'gold standard' to assist in GH testing as all parameters have arbitrary cut-off levels and low accuracy [3–6]. In addition, neither serum insulin-like growth factor 1 (IGF-1) nor IGF-binding protein 3 (IGFBP-3) measurements are by themselves of predictive value [7]. However, a sound clinical diagnosis is crucial before any treatment so the decision-making processes must be viewed in the total clinical setting. Here we see false-positively tested children diagnosed as 'partial GHD', whereas the false-negative ones are labeled as ISS. But it is possible that in both cases the short stature arises because of a diminished GH secretion for a given GH sensitivity.

However, as we know that recombinant human GH (rhGH) leads in almost all children to an increase of height velocity in the first year, which tapers off in the following years, we have to assume that in the vast majority of those children either the endogenous GH secretion is suboptimal for normal growth, or that the GH sensitivity/responsiveness can be increased by the administration of either physiological or pharmacological doses of rhGH. Moreover, even at the level of GH, many genes are involved in controlling growth regulation. As these disturbances can be divided into two gross categories, namely alterations causing subnormal GH secretion and/or subnormal GH sensitivity/responsiveness, the main aim of this review is to focus on genes involved in growth regulation leading to short stature caused by an alteration of GH insensitivity/GH responsiveness. In these clinical circumstances, individually adapted GH replacement therapy may help to increase height velocity and eventually final height [8] although as discussed in Chapter 2 the two may not be synonymous.

Pharmacogenomics of Growth

Pharmacogenetics (impact of one gene) and pharmacogenomics (impact of several genes, genome) is the study how a person's gene/genome can influence his/

her response to medication [9]. Based on the research in the field of GHD, we know that GHD is quite heterogeneous in terms of etiology as well as age at diagnosis and that improvement in adult height (over and above that of the untreated state) is the major aim of treating GHD children with rhGH [10]. The final height attained as a result of intervention is influenced, in part, by the dose, injection frequency and duration of rhGH therapy. Despite optimization of these factors, a proportion of GHD patients do not reach their target height [10, 11].

A number of mathematical models for predicting growth and final outcome have been proposed enabling the clinician to 'personalize' the growth-promoting therapy on the grounds of efficacy and economy [12] although whether the factors identified really impact on response has not been tested in formal randomized control trials. There are several problems with these types of models:

(1) Although prediction models are useful to give an average effect, they are not individualizable.

(2) They often only focus on one outcome, usually short-term growth, whereas interest may be more centered on final height. The two need not necessarily be related and the factors that influence response in the first year of treatment may differ totally from those that lead to prediction of the individual's final height.

(3) Very few prediction models have been constructed from an a priori hypothesis and care needs to be taken that there has been no interference from other factors accompanying the disease that might affect prognosis. The problem is that importance can be ascribed to factors that are merely 'markers' for other factors of real importance. Examples of this can be seen in models which demonstrate that individuals who are extremely short, growing very poorly and whose heights are subsequently further away from their genetic height respond best to treatment. All these factors are simply a marker of 'how bad the disease is' and could perhaps be more easily summarized by a similar single factor that actually describes the severity of the condition.

(4) Rules derived from one data set may reflect associations that have occurred by chance and often result from overfitting of the data.

(5) There is always the possibility that the predictors are idiosyncratic to the population, the setting, to the clinicians or to other aspects of the original study.

The identification of these parameters is understandable as they are easy to measure as opposed underlying genetic and epigenetic factors that might explain this individual variability of GH response. However, with an increased understanding of the factors involved in human growth, these genetic and epigenetic factors may become more important and accessible. For example, based on animal knockouts and human mutational analyses the most factors/genes affecting IGF generation as well as the structure of the growth plate deserve consideration [13].

In future, therefore, the use of specifically designed and personalized prediction models may well facilitate the decision about whether the growth response to a given therapy (rhGH; rhIGF-1) in an individual child is appropriate or not [14, 15]. Based on modifiable (start of treatment, optimal dose of treatment, etc.) and non-modifiable (start heights SDS, bone age, target height SDS, etc.) variables including genetic/genomic variants as well as phenotype-genotype relationship, the realistic growth potential will be calculated. With regard to GH treatment, pharmacogenomics may play, therefore, a major role in the individual response to therapy, at least in the 'sub'-group of subjects not following the current prediction models.

GH Insensitivity (GHI)

The classic phenotype of severe growth failure associated with elevated serum GH concentrations was first described by Laron et al. [16] and can be classified as primary IGF-1 deficiency [17, 18]. Severe forms of primary IGF-1 deficiency have been observed with molecular defects involving the GH receptor (GHR), the GHR cascade and IGF-1/acid-labile subunit/IGFBP-3 ternary complex as well as the type 1 IGF receptor. Based on molecular data, defects anywhere along the pathway from GH binding to its receptor to the IGF-1 action at the growth plate may contribute to postnatal growth failure [17, 18]. This does not include the complex biology of the growth plate manifest at present only in our understanding of the importance of the fibroblast growth factor receptor-3 in achondroplasia. Although we are accustomed to gene deletion/mutation leading to disease, there is clear evidence that either a gene haploinsufficiency, a common polymorphism and/or partially disturbed signaling pathway may impact on susceptibility to disease or modification of treatment response in a number of ways.

GH Receptor

Growth defects result from rare molecular defects including exon deletions or mutations (nonsense, frameshift, missense) of the GHR [19] are unusual in humans. Three types of *GHR* mutations either affecting expression, activation or signaling are specifically highlighted in this review, but consideration is also given to heterozygous forms of *GHR* mutations as well as polymorphism within *GHR* gene possibly affecting growth as exemplars of what has already been discussed.

Expression Failure [20, 21]
Inherited GHI is a heterogeneous disorder that is often caused by mutations in the coding exons or flanking intronic sequences of the *GHR* gene. In 4 children with GHI, Metherell et al. [21] described a novel point mutation that led to activation

of an intronic pseudoexon resulting in inclusion of an additional 108 nt between exons 6 and 7 in the majority of GHR transcripts. This mutation lies within the pseudoexon [A(-1)→G(-1) at the 5′ pseudoexon splice site] and, under in vitro splicing conditions, results in inclusion of the mutant pseudoexon, whereas the wild-type pseudoexon is skipped. The presence of the pseudoexon results in inclusion of an additional 36-amino-acid sequence in a region of the receptor previously alleged to be involved in homodimerization, which may be essential for signal transduction [21]. Based on functional studies, Maamra et al. [20] have shown that this elongated GHR remained trapped around the nucleus and is therefore poorly expressed at the cell membrane, reflecting a trafficking defect and, thus, reduced downstream signaling. These properties may explain the relatively mild phenotype of these subjects.

Activation Failure
The substitution of histidine for aspartate 152 (D152H GHR) has been described in a context of familial GH resistance [22]. Originally, this mutation was suggested to interfere with receptor homodimerization because it was found to abolish homodimerization of GHBPs. But later, Waters et al. [23, 24] showed that GHR is constitutively homodimerized at the cell membrane, which presumably involves contacts between transmembrane or juxtamembrane domains of each receptor chain. The activation process mediated by the ligand is assumed to involve a conformational change such as relative rotation of upper and lower domains of the receptor. Based on this model, D152H GHR may interfere with this conformational change.

Signaling Failure
In a short statured family, Ross et al. [25, 26] documented a heterozygous expression of a severely truncated GHR mutant resulting from a mutation at the splice acceptor site of exon 3. A GHR short of the cytoplasmic domain would be devoid of any signaling capacity and because internalization is also impossible, such truncated receptors accumulate at the membrane and act as dominant-negative forms [25]. Less dramatically truncated, homozygous/compound heterozygous GHR leading to GHI have been reported [27, 28]. The truncations of the GHR were after residue 449 (nonsense sequence of residues 424–449) and 581 (nonsense sequence of residues 560–581), respectively. In both cases, STAT5 activation, a component of subsequent post-GHR signaling, was drastically impaired [29].

Heterozygosities for GHR Mutations
There are several reports presenting data on heterozygous *GHR* gene defects possibly leading to short stature [30–33]. Defects in the *GHR* gene were present at a modest frequency (approx. 30%) in persons who were selected for short

stature (height SDS <–2), low GH binding protein – and IGF-1 concentrations and poor height velocity. The frequency is much lower in short subjects (approx. 2%) presenting with adequate GH concentrations. However, Rosenbloom et al. [34] found minimal or no effect on stature analyzing subjects presenting with heterozygous *GHR* gene mutations. Further, Johnston et al. [35] studied the intracellular signaling domain of the GHR in children with ISS and concluded that ISS is not related to heterozygous and/or dominant-negative *GHR* variants. Interestingly, in the study by Woods et al. [33] focusing on phenotype-genotype relationships the mean adult heights of both mothers and fathers were reduced when compared to British standards from 1958. These standards, however, are not ideal for comparison in view of the diverse ethnic origin of the patients, but this finding suggests the possibility of a heterozygote effect. Furthermore, among the 19 families in whom a homozygous *GHR* defect in the affected children was found, the height deficit between the two parents was generally not uniform, suggesting that other genes may be influencing the magnitude of heterozygote effect in each individual. Further, in this same report, certain *GHR* gene mutations producing the GHBP-positive phenotype, in which GH binding is normal, were described. These forms are more likely to act in a semi-dominant manner. This is because dimeric GH binding is a prerequisite for GHR activation so that mutant receptors that bind GH normally could dimerize with wild-type GHR, reducing the number of active wild-type homodimers. This hypothesis was tested, but the expected decrease in mean parental height SDS between the GHBP-positive group when compared with those in the GHBP-negative group could not be found [33].

Polymorphism of the GHR
While a mutation can change the amino acid sequence and influence the transcript function, a polymorphism is not expected to cause a major change in protein function. A polymorphism is defined as a DNA sequence variant that occurs in at least 1% of the population [36]. Polymorphisms of the *GHR* gene have been reported in the general population and have been described in exons 3, 6 and 10 [31]. While the latter two are classical single nucleotide polymorphisms, the polymorphism in exon 3 is an unusual one, leading to retention (full-length, fl; GHRfl) or deletion of exon 3 (d3; GHRd3), which encodes a 22-amino-acid residue sequence in the extracellular domain [37, 38]. GHRd3 has recently been associated with the degree of height increase in response to GH replacement in children born short for gestational age (SGA), in those with ISS, and in a GHD population [39, 40]. Patients with at least one GHRd3 allele (GHRfl/GHRd3; GHRd3/GHRd3) had a significantly better first year response leading to an improved adult height on rhGH treatment than patients with homozygosity for GHRfl [39]. However, reported studies are not all consistent which may reflect

differing populations and conditions and the high probability of false-positive results arising from post-hoc analysis particularly in small sample size populations [41–49].

In a recent study we analyzed in a total of 186 subjects the impact of GHR genotypes (GHRd3/d3; GHRd3/fl; GHRfl/fl) on growth response to rhGH replacement therapy in two groups of patients (group A: mean rhGH dose: 26.5 µg/kg/day, n = 104; group B: 36.5 µg/kg/day, n = 82) suffering from severe idiopathic isolated GHD (mean maximal GH peak on GH stimulation tests: 0.6 and 1.3 ng/ml, respectively) [12]. Importantly, these patients were followed up to final height and were individually analyzed for the first 4 years on rhGH replacement therapy. In contrast to previous reports, the subjects with the GHRd3/d3 and GHRd3/fl were not pooled but analyzed separately [39, 45, 47]. Overall, in the subjects presenting with either the GHRd3/d3 or GHRd3/fl genotype, a significantly better response (height velocity) to the replacement therapy during the first 2 years of therapy was noted, although in the third and fourth year of therapy, this improved height velocity was observed in those patients with the GHRfl/fl genotype. Further, during the first 2 years on therapy a GHRd3 allele-dependent effect on height was found in study A (r = 0.82), which could not be reported in study B, where a higher rhGH dose was used. However, at final height the effect of rhGH treatment was identical irrespective of the specific GHR genotype. Comparing the difference between final adult height SDS and the midparental target height SDS in our severely GHD patients with the large cohort of Caucasian GHD children recruited from the KIGS database, no difference was noted [50]. Similar data were previously published by the Genentech Growth Study Group underlining the effectiveness of the rhGH treatment used in our studies [10]. From these data reporting final height it can be concluded that in severe GHD subjects, the presence or absence of the GHRd3 allele has no impact on either baseline phenotype or final height, although a difference in response (height velocity a different parameter) to rhGH replacement therapy during the first years may be observed depending on the genotype.

When comparing these findings obtained from patients with severe GHD with the previous studies focusing on GHR allele genotype and response to rhGH replacement therapy, the patients, the individual conditions, their related growth disorder as well as the rhGH doses used have to be carefully analyzed [41–49]. In the first report, for instance, Dos Santos et al. [40] studied patients with either SGA or ISS so that in effect patients with normal GH secretion were treated with supraphysiological rhGH doses. Besides the various conditions studied it is also possible that the differences between the studies reported so far represent the problems of sample size. False-positive findings are more likely with small samples sizes and for quantitative trait loci phenotypic variations tend to be overestimated with small sample sizes [51, 52]. Only large-scale studies of well-defined conditions or pooling the data sets will help to resolve these statistical issues.

Bearing in mind that rhGH replacement and/or therapy in any subject does not result in a constant increase of height velocity over the whole duration of treatment underlines the fact that neither GH responsiveness nor GH sensitivity is constant. Dose-response relationship may vary in every specific condition with or without any underlying growth disorder. Even for a specific condition and dose-response curve, an increase of rhGH dose may well itself affect GHR sensitivity, signaling and thus response to treatment. This fact is well established in any rhGH-treated child whose response changes after the first years on treatment [12]. Moreover, the dose response of rhGH differs according to the condition that is treated. For instance, children with GHD, Turner syndrome, SGA or ISS respond differently (and the rhGH dose is adapted accordingly) but a change in GH sensitivity with treatment remains a variable. Thus, the positive effect resulting from the GHRd3 genotype may well be downregulated and/or altered when supraphysiological doses of rhGH are given. This hypothesis could explain why severe GHD subjects might present a rhGH dose-dependent GHR genotype-related effect on growth response with an apparent plateau at around 26 µg/kg/day (16 IU/m^2/week), whereas this effect disappears with higher rhGH doses that lie further up on the dose-response curve for this condition. Similarly, SGA children treated with higher doses of rhGH showed no difference in growth response according to the GHR exon-3 genotype in contrast to SGA children treated with lower doses [40, 41]. Duration of GH treatment also seems to play a major role in defining the sensitivity to the GHR genotype, in addition to the etiology of the short stature and its severity.

In summary, focusing on patients with severe IGHD, we observed a GHRd3 allele dose-dependent effect in subjects treated with rhGH during the first 2 years, with significantly better responses depending on GHRd3/d3:GHRd3/fl genotype, although no difference was observed at final height. Taking all the studies focusing on isolated GHD into account it becomes clear that the final impact of the GHR genotypes on the rhGH response is minimal [39, 43, 45, 47]. The same finding seems to be true for children treated suffering from SGA [48, 53]. Given the importance of the response to attainment of final height it may, nevertheless, well be an additional variable having some impact on growth and GH sensitivity, at least at the beginning of rhGH treatment. These findings are also supported by a clinical case report of a child with GHI syndrome, caused by a compound heterozygosity of *GHR* gene mutation. Father (mutation in exon 4 leading to a stop codon) and mother (mutation in exon 3 leading to a stop codon) carrying either GHRfl (father) or GHRd3 (mother) were of normal stature. Therefore the authors concluded that a single copy of either GHRfl or GHRd3 is sufficient for normal growth [38].

Altered GHR Signaling

Defects in the GHR signaling pathway deserve further consideration given the critical role played in rodent growth. Each subunit of the dimeric GHR associates non-covalently (through its box one motif) with a molecule of cytosolic Janus-family tyrosine kinase 2 (JAK2). Following binding of one GH molecule, the dimeric GHR undergoes conformational changes that induce transphosphorylation of JAK2 and initiation of GHR signaling. Ligand-activated JAK2 phosphorylates multiple tyrosines on the intracellular domain of the GHR, which then serve as docking sites for cytosolic components of at least three distinct signaling pathways: the signal transducer and activator of transcription (STAT), the mitogen-activated protein kinase (MAPK) and the phosphoinositide 3-kinase (PI3K) pathways. A proline-rich region in the intracellular N-terminal domain (ND; residues 279–286 in the box one motif) seems to be required for JAK2 and MAPK activation, as well as for phosphorylation of STAT1 and STAT3; tyrosines in the C-terminal portion of the intracellular domain are essential for STAT5 activation. These signaling cascades culminate in the regulation of multiple genes [24, 29, 54]. Our understanding of each pathway for GH-promoted functions has been based predominantly on studies employing rodent models and reconstitution systems. Valuable insights can be gained from identification of defective intracellular components of GH signaling in human disorders but the pleiotropic effects of such defects often complicate characterization. To date, only a limited number of mutations of intracellular GH signaling components have been demonstrated to be convincingly associated with growth retardation.

STAT5b
The identification of, to date, 10 subjects suffering from severe growth failure associated with homozygosity for mutations of the *STAT5b* gene has confirmed the central role of STAT5b in the GH-induced IGF-1 expression and in mammalian postnatal growth [29]. Therefore, in this circumstance GHI is most severe.

Other Defects Leading to GHI Syndromes
So far, no reports of growth failure caused by *JAK2* gene alterations have been published. Further, impaired STAT3 activation has been reported in ISS [55]. However, no mutations were identified, but following GH therapy increased IGF-1 levels in addition to an increase in height velocity was observed.

PTPN11-Gene
Noonan syndrome (OMIM: #163950) is an autosomal dominant dysmorphic syndrome characterized by hypertelorism, a downward eyeslant, and low-set posteriorly rotated ears. Other features include short stature, a short neck with webbing

or redundancy of skin, cardiac anomalies, epicanthic folds, deafness, motor delay, and a bleeding diathesis. Approximately 50% of cases have been associated with gain-of-function mutations of *PTPN11*, the gene encoding the non-receptor-type protein tyrosine phosphatase src homology region 2-domain phosphatase-2 (SHP-2) [56]. This tyrosine phosphatase is involved in intracellular signaling for a variety of hormones, growth factors and cytokines. Activated SHP-2 has been implicated as a negative regulator of GH signaling, and mutations of the SHP-2-binding site in the GHR prolongs GH-promoted tyrosyl phosphorylation of the GHR, JAK2 and STAT5b [57, 58].

Recent reports have indicated more severe stature impairment in Noonan patients with mutations in *PTPN11*, as well as lower serum IGF-1 and IGFBP-3 concentrations, higher GH concentrations and a more modest response to GH therapy, consistent with mild GH resistance [59, 60]. Heterozygous mutations in the *KRAS* gene, downstream effector of SHP-2, were also recently reported to be associated with Noonan syndrome but the impact of such mutations on components of the GHR signaling pathway is not clear [61]. Another potential site for molecular defects of GHR signaling is suppressor of cytokine signaling-2 (*SOCS2*), which is involved in the negative regulation of cytokine action through the inhibition of JAKs and STATs [62, 63].

Growth Responsiveness to GH Therapy, the Impact of the Growth Plate

Studies are already in progress to assess both proteomic and genomic biomarkers for the evaluation and management of short stature and for the assessment of responsiveness to GH therapy. While mutations and polymorphisms of known genes involved in the GH/IGF-1 axis have been identified in various forms of short stature, attention should also be drawn to non-GH/IGF-related factors such as SHOX (see below), fibroblast growth factor receptor-3 [64], the C-type natriuretic peptide receptor 2 (NPR2) [65] known to affect linear growth as well as growth plate structure and impact, therefore, on the effect of GH/IGF axis as well. In addition to other factors yet to be identified, it is also very likely that epigenetic factors will provide important insights into the evaluation of short stature.

Longitudinal bone growth occurs rapidly in early life and slows down and eventually ceases at the end of puberty. This decline in growth rate is due primarily to a decrease in the rate of chondrocyte proliferation and is accompanied by a structural change in growth plate cartilage. Although in humans the age-dependent decline in growth rate is interrupted by a brief period of growth acceleration, which peaks during early to mid puberty and may partly be induced by estrogen increasing the activity of GH/IGF-1 axis [66, 67], this programmed senescence appears not to be caused by a systemic mechanism but rather by a mechanism intrinsic to the growth plate itself. Based on data focusing on growth plate senescence and hormonal impact it becomes clear that the enhanced growth responsiveness of GH at the young age is unlikely to be mediated by IGF-1, but appears to reside within the growth plate itself [68–70]. Any disorder at the level of growth plate is bound to result in short stature and may present with different growth responsiveness and, therefore, a given GH therapy has to be correspondingly personalized [71]. In addition, for these non-GHD conditions, GH is used in a pharmacological manner rather than as physiologic replacement, and, therefore, the GH dose is rather high in order to improve height velocity and increase final height [72].

SHOX (Short Stature HOmeoboX Containing Gene; OMIM: #312865)
The *SHOX* gene was discovered in 1997 during the search for genes underlying the short stature of Turner syndrome and is located in the pseudoautosomal regions at the distal ends of the X and Y chromosome, at positions Xp22.3 and Yp11.3. The gene encodes a homeodomain transcription factor responsible for a significant proportion of long bone growth [73]. In addition to its role in explaining the growth deficit in Turner syndrome, SHOX haploinsufficiency is also the primary cause of short stature in 50–70% of individuals who have Leri-Weill dyschondrosteosis (LWD, OMIM: #127300) and in about 2–15% of ISS [72, 74, 75].

Clinically, SHOX deficiency is associated with a broad spectrum of phenotypic effects, ranging from short stature without dysmorphic signs to profound mesomelic skeletal dysplasia, a form of short stature characterized by disproportionate shortening of the middle (mesial) segments of the upper as well as lower limbs [76]. There is strong evidence that SHOX is a major mediator of linear growth; first, it is expressed in the developing skeleton during fetal life and is specifically expressed in bone marrow fibroblasts and proliferating hypertrophic chondrocytes; second, deficiency of SHOX at the growth plate is associated with marked disorganization of chondrocyte proliferation [77], and third, there is a dose-dependent association between the number of active copies of the SHOX gene and height [78].

Conclusions

The concept that genetic variation contributes in general to variability in disease phenotypes and therefore in drug responses is widely accepted and validated in many research settings. Therefore, the identification of an association between a clinical phenotype, such as short stature, and a genetic variant, for instance within the *GHR* gene, or a set of genetic variants is an increasing theme also in the field of pediatric endocrinology. Although many such associations were not reproduced in subsequent studies and there are many challenges to overcome in the implementation of pharmacogenetic as well as pharmacogenomic vision in clinical practice, potential solutions are also evolving rapidly and may help to individualize GH treatment based upon well-defined GH responsiveness and careful genetic analysis.

References

1 Prader A, Largo RH, Molinari L, Issler C: Physical growth of Swiss children from birth to 20 years of age. First Zurich longitudinal study of growth and development. Helv Paediatr Acta Suppl 1989;52:1–125.

2 Mullis PE: Genetic control of growth. Eur J Endocrinol 2005;152:11–31.

3 Hindmarsh P, Smith PJ, Brook CG, Matthews DR: The relationship between height velocity and growth hormone secretion in short prepubertal children. Clin Endocrinol (Oxf) 1987;27:581–591.

4 Dammacco F, Boghen MF, Camanni F, Cappa M, Ferrari C, Ghigo E, Giordano G, Loche S, Minuto F, Mucci M, et al: Somatotropic function in short stature: evaluation by integrated auxological and hormonal indices in 214 children. The Italian Collaborative Group of Neuroendocrinology. J Clin Endocrinol Metab 1993;77:68–72.

5 Rogol AD, Blethen SL, Sy JP, Veldhuis JD: Do growth hormone (GH) serial sampling, insulin-like growth factor-1 (IGF-1) or auxological measurements have an advantage over GH stimulation testing in predicting the linear growth response to GH therapy? Clin Endocrinol (Oxf) 2003;58:229–237.

6 Hindmarsh P: Endocrine assessment and principles of endocrine testing; in Kelnar C, Savage M, Saenger P, Cowell C (eds): Growth disorders. London, Hodder Arnold, 2007, pp 219–229.

7 Mitchell H, Dattani MT, Nanduri V, Hindmarsh PC, Preece MA, Brook CG: Failure of IGF-1 and IGFBP-3 to diagnose growth hormone insufficiency. Arch Dis Child 1999;80:443–447.

8 Rosenfeld RG: Pharmacogenomics and pharmacoproteomics in the evaluation and management of short stature. Eur J Endocrinol 2007;157(suppl 1):S27–S31.

9 Thomas FJ, McLeod HL, Watters JW: Pharmacogenomics: the influence of genomic variation on drug response. Curr Top Med Chem 2004;4:1399–1409.

10 Blethen SL, Baptista J, Kuntze J, Foley T, LaFranchi S, Johanson A: Adult height in growth hormone (GH)-deficient children treated with biosynthetic GH. The Genentech Growth Study Group. J Clin Endocrinol Metab 1997;82:418–420.

11 Ranke MB, Lindberg A, Chatelain P, Wilton P, Cutfield W, Albertsson-Wikland K, Price DA: Derivation and validation of a mathematical model for predicting the response to exogenous recombinant human growth hormone (GH) in prepubertal children with idiopathic GH deficiency. KIGS International Board. Kabi Pharmacia International Growth Study. J Clin Endocrinol Metab 1999;84:1174–1183.

12 Raz B, Janner M, Petkovic V, Lochmatter D, Eble A, Dattani MT, Hindmarsh PC, Fluck CE, Mullis PE: Influence of growth hormone (GH) receptor deletion of exon 3 and full-length isoforms on GH response and final height in patients with severe GH deficiency. J Clin Endocrinol Metab 2008;93:974–980.

13 Rosenfeld RG: The pharmacogenomics of human growth. J Clin Endocrinol Metab 2006;91:795–796.

14 Ranke MB, Lindberg A, Chatelain P, Wilton P, Price DA, Albertsson-Wikland K: The potential of prediction models based on data from KIGS as tools to measure responsiveness to growth hormone. Pharmacia International Growth Database. Horm Res 2001;55(suppl 2):44–48.

15 Weinshilboum R, Wang L: Pharmacogenomics: bench to bedside. Nat Rev Drug Discov 2004;3:739–748.

16 Laron Z, Pertzelan A, Mannheimer S: Genetic pituitary dwarfism with high serum concentration of growth hormone – a new inborn error of metabolism? Isr J Med Sci 1966;2:152–155.

17 Rosenfeld RG, Hwa V: Toward a molecular basis for idiopathic short stature. J Clin Endocrinol Metab 2004;89:1066–1067.

18   Rosenfeld RG, Kofoed E, Little B, Woods K, Buckway C, Pratt K, Hwa V: Growth hormone insensitivity resulting from post-GH receptor defects. Growth Horm IGF Res 2004;14(suppl A):S3-5–S38.

19   Bougneres P, Goffin V: The growth hormone receptor in growth. Endocrinol Metab Clin North Am 2007;36:1–16.

20   Maamra M, Milward A, Esfahani HZ, Abbott LP, Metherell LA, Savage MO, Clark AJ, Ross RJ: A 36 residues insertion in the dimerization domain of the growth hormone receptor results in defective trafficking rather than impaired signaling. J Endocrinol 2006;188:251–261.

21   Metherell LA, Akker SA, Munroe PB, Rose SJ, Caulfield M, Savage MO, Chew SL, Clark AJ: Pseudoexon activation as a novel mechanism for disease resulting in atypical growth-hormone insensitivity. Am J Hum Genet 2001;69:641–646.

22   Duquesnoy P, Sobrier ML, Duriez B, Dastot F, Buchanan CR, Savage MO, Preece MA, Craescu CT, Blouquit Y, Goossens M, et al: A single amino acid substitution in the exoplasmic domain of the human growth hormone (GH) receptor confers familial GH resistance (Laron syndrome) with positive GH-binding activity by abolishing receptor homodimerization. EMBO J 1994;13:1386–1395.

23   Brown RJ, Adams JJ, Pelekanos RA, Wan Y, McKinstry WJ, Palethorpe K, Seeber RM, Monks TA, Eidne KA, Parker MW, Waters MJ: Model for growth hormone receptor activation based on subunit rotation within a receptor dimer. Nat Struct Mol Biol 2005;12:814–821.

24   Waters MJ, Hoang HN, Fairlie DP, Pelekanos RA, Brown RJ: New insights into growth hormone action. J Mol Endocrinol 2006;36:1–7.

25   Ross RJ, Esposito N, Shen XY, Von Laue S, Chew SL, Dobson PR, Postel-Vinay MC, Finidori J: A short isoform of the human growth hormone receptor functions as a dominant negative inhibitor of the full-length receptor and generates large amounts of binding protein. Mol Endocrinol 1997;11:265–273.

26   Ayling RM, Ross R, Towner P, Von Laue S, Finidori J, Moutoussamy S, Buchanan CR, Clayton PE, Norman MR: A dominant-negative mutation of the growth hormone receptor causes familial short stature. Nat Genet 1997;16:13–14.

27   Milward A, Metherell L, Maamra M, Barahona MJ, Wilkinson IR, Camacho-Hubner C, Savage MO, Bidlingmaier CM, Clark AJ, Ross RJ, Webb SM: Growth hormone (GH) insensitivity syndrome due to a GH receptor truncated after Box1, resulting in isolated failure of STAT5 signal transduction. J Clin Endocrinol Metab 2004;89:1259–1266.

28   Tiulpakov A, Rubtsov P, Dedov I, Peterkova V, Bezlepkina O, Chrousos GP, Hochberg Z: A novel C-terminal growth hormone receptor (GHR) mutation results in impaired GHR-STAT5 but normal STAT-3 signaling. J Clin Endocrinol Metab 2005;90:542–547.

29   Rosenfeld RG, Belgorosky A, Camacho-Hubner C, Savage MO, Wit JM, Hwa V: Defects in growth hormone receptor signaling. Trends Endocrinol Metab 2007;18:134–141.

30   Goddard AD, Dowd P, Chernausek S, Geffner M, Gertner J, Hintz R, Hopwood N, Kaplan S, Plotnick L, Rogol A, Rosenfield R, Saenger P, Mauras N, Hershkopf R, Angulo M, Attie K: Partial growth-hormone insensitivity: the role of growth-hormone receptor mutations in idiopathic short stature. J Pediatr 1997;131:S51–55.

31   Goddard AD, Covello R, Luoh SM, Clackson T, Attie KM, Gesundheit N, Rundle AC, Wells JA, Carlsson LM: Mutations of the growth hormone receptor in children with idiopathic short stature. The Growth Hormone Insensitivity Study Group. N Engl J Med 1995;333:1093–1098.

32   Sanchez JE, Perera E, Baumbach L, Cleveland WW: Growth hormone receptor mutations in children with idiopathic short stature. J Clin Endocrinol Metab 1998;83:4079–4083.

33   Woods KA, Dastot F, Preece MA, Clark AJ, Postel-Vinay MC, Chatelain PG, Ranke MB, Rosenfeld RG, Amselem S, Savage MO: Phenotype:genotype relationships in growth hormone insensitivity syndrome. J Clin Endocrinol Metab 1997;82:3529–3535.

34   Rosenbloom AL, Guevara-Aguirre J, Rosenfeld RG, Pollock BH: Growth in growth hormone insensitivity. Trends Endocrinol Metab 1994;5:296–303.

35   Johnston LB, Pashankar F, Camacho-Hubner C, Savage MO, Clark AJ: Analysis of the intracellular signalling domain of the human growth hormone receptor in children with idiopathic short stature. Clin Endocrinol (Oxf) 2000;52:463–469.

36 Roden DM, Altman RB, Benowitz NL, Flockhart DA, Giacomini KM, Johnson JA, Krauss RM, McLeod HL, Ratain MJ, Relling MV, Ring HZ, Shuldiner AR, Weinshilboum RM, Weiss ST: Pharmacogenomics: challenges and opportunities. Ann Intern Med 2006;145:749–757.

37 Pantel J, Machinis K, Sobrier ML, Duquesnoy P, Goossens M, Amselem S: Species-specific alternative splice mimicry at the growth hormone receptor locus revealed by the lineage of retroelements during primate evolution. J Biol Chem 2000;275:18664–18669.

38 Pantel J, Grulich-Henn J, Bettendorf M, Strasburger CJ, Heinrich U, Amselem S: Heterozygous nonsense mutation in exon 3 of the growth hormone receptor (GHR) in severe GH insensitivity (Laron syndrome) and the issue of the origin and function of the GHRd3 isoform. J Clin Endocrinol Metab 2003;88:1705–1710.

39 Jorge AA, Marchisotti FG, Montenegro LR, Carvalho LR, Mendonca BB, Arnhold IJ: Growth hormone (GH) pharmacogenetics: influence of GH receptor exon 3 retention or deletion on first-year growth response and final height in patients with severe GH deficiency. J Clin Endocrinol Metab 2006;91:1076–1080.

40 Dos Santos C, Essioux L, Teinturier C, Tauber M, Goffin V, Bougneres P: A common polymorphism of the growth hormone receptor is associated with increased responsiveness to growth hormone. Nat Genet 2004;36:720–724.

41 Carrascosa A, Esteban C, Espadero R, Fernandez-Cancio M, Andaluz P, Clemente M, Audi L, Wollmann H, Fryklund L, Parodi L: The d3/fl-growth hormone (GH) receptor polymorphism does not influence the effect of GH treatment (66 µg/kg per day) or the spontaneous growth in short non-GH-deficient small-for-gestational-age children: results from a two-year controlled prospective study in 170 Spanish patients. J Clin Endocrinol Metab 2006;91:3281–3286.

42 Binder G, Baur F, Schweizer R, Ranke MB: The d3-growth hormone (GH) receptor polymorphism is associated with increased responsiveness to GH in Turner syndrome and short small-for-gestational-age children. J Clin Endocrinol Metab 2006;91:659–664.

43 Pilotta A, Mella P, Filisetti M, Felappi B, Prandi E, Parrinello G, Notarangelo LD, Buzi F: Common polymorphisms of the growth hormone (GH) receptor do not correlate with the growth response to exogenous recombinant human GH in GH-deficient children. J Clin Endocrinol Metab 2006;91:1178–1180.

44 Audi L, Esteban C, Carrascosa A, Espadero R, Perez-Arroyo A, Arjona R, Clemente M, Wollmann H, Fryklund L, Parodi LA: Exon 3-deleted/full-length growth hormone receptor polymorphism genotype frequencies in Spanish short small-for-gestational-age (SGA) children and adolescents (n = 247) and in an adult control population (n = 289) show increased fl/fl in short SGA. J Clin Endocrinol Metab 2006;91:5038–5043.

45 Blum WF, Machinis K, Shavrikova EP, Keller A, Stobbe H, Pfaeffle RW, Amselem S: The growth response to growth hormone (GH) treatment in children with isolated GH deficiency is independent of the presence of the exon 3-minus isoform of the GH receptor. J Clin Endocrinol Metab 2006;91:4171–4174.

46 Audi L, Carrascosa A, Esteban C, Fernandez-Cancio M, Andaluz P, Yeste D, Espadero R, Granada ML, Wollmann H, Fryklund L: The exon 3-deleted/full-length growth hormone receptor polymorphism does not influence the effect of puberty or growth hormone therapy on glucose homeostasis in short non-growth hormone-deficient small-for-gestational-age children: results from a two-year controlled prospective study. J Clin Endocrinol Metab 2008;93:2709–2715.

47 Van der Klaauw AA, van der Straaten T, Baak-Pablo R, Biermasz NR, Guchelaar HJ, Pereira AM, Smit JW, Romijn JA: Influence of the d3-growth hormone (GH) receptor isoform on short-term and long-term treatment response to GH replacement in GH-deficient adults. J Clin Endocrinol Metab 2008;93:2828–2834.

48 Carrascosa A, Audi L, Fernandez-Cancio M, Esteban C, Andaluz P, Vilaro E, Clemente M, Yeste D, Albisu MA, Gussinye M: The exon 3-deleted/full-length growth hormone receptor polymorphism did not influence growth response to growth hormone therapy over two years in prepubertal short children born at term with adequate weight and length for gestational age. J Clin Endocrinol Metab 2008;93:764–770.

49 Toyoshima MT, Castroneves LA, Costalonga EF, Mendonca BB, Arnhold IJ, Jorge AA: Exon 3-deleted genotype of growth hormone receptor (GHRd3) positively influences IGF-1 increase at generation test in children with idiopathic short stature. Clin Endocrinol (Oxf) 2007;67:500–504.

50 Reiter EO, Price DA, Wilton P, Albertsson-Wikland K, Ranke MB: Effect of growth hormone (GH) treatment on the near-final height of 1258 patients with idiopathic GH deficiency: Analysis of a large international database. J Clin Endocrinol Metab 2006;91:2047–2054.

51 Xu S: Theoretical basis of the Beavis effect. Genetics 2003;165:2259–2268.

52 Beavis W: QTL analyses: power, precision and accuracy; in Paterson A (ed): Molecular Dissection of Complex Traits. Boca Raton, CRC Press, 1998, pp 145–161.

53 Carrascosa A, Audi L, Esteban C, Fernandez-Cancio M, Andaluz P, Gussinye M, Clemente M, Yeste D, Albisu MA: Growth hormone (GH) dose, but not exon 3-deleted/full-length GH receptor polymorphism genotypes, influences growth response to two-year GH therapy in short small-for-gestational-age children. J Clin Endocrinol Metab 2008;93:147–153.

54 Zhu T, Goh EL, Graichen R, Ling L, Lobie PE: Signal transduction via the growth hormone receptor. Cell Signal 2001;13:599–616.

55 Rojas-Gil AP, Ziros PG, Diaz L, Kletsas D, Basdra EK, Alexandrides TK, Zadik Z, Frank SJ, Papathanassopoulou V, Beratis NG, Papavassiliou AG, Spiliotis BE: Growth hormone/JAK-STAT axis signal-transduction defect. A novel treatable cause of growth failure: FEBS J 2006;273:3454–3466.

56 Tartaglia M, Kalidas K, Shaw A, Song X, Musat DL, van der Burgt I, Brunner HG, Bertola DR, Crosby A, Ion A, Kucherlapati RS, Jeffery S, Patton MA, Gelb BD: PTPN11 mutations in Noonan syndrome: molecular spectrum, genotype-phenotype correlation, and phenotypic heterogeneity. Am J Hum Genet 2002;70:1555–1563.

57 Carter-Su C, Rui L, Stofega MR: SH2-B and SIRP: JAK2 binding proteins that modulate the actions of growth hormone. Recent Prog Horm Res 2000; 55:293–311.

58 Stofega MR, Herrington J, Billestrup N, Carter-Su C: Mutation of the SHP-2 binding site in growth hormone (GH) receptor prolongs GH-promoted tyrosyl phosphorylation of GH receptor, JAK2, and STAT5b. Mol Endocrinol 2000;14:1338–1350.

59 Binder G, Neuer K, Ranke MB, Wittekindt NE: Ptpn11 mutations are associated with mild growth hormone resistance in individuals with Noonan syndrome. J Clin Endocrinol Metab 2005;90:5377–5381.

60 Limal JM, Parfait B, Cabrol S, Bonnet D, Leheup B, Lyonnet S, Vidaud M, Le Bouc Y: Noonan syndrome: relationships between genotype, growth, and growth factors. J Clin Endocrinol Metab 2006;91:300–306.

61 Schubbert S, Zenker M, Rowe SL, Boll S, Klein C, Bollag G, van der Burgt I, Musante L, Kalscheuer V, Wehner LE, Nguyen H, West B, Zhang KY, Sistermans E, Rauch A, Niemeyer CM, Shannon K, Kratz CP: Germline KRAS mutations cause Noonan syndrome. Nat Genet 2006;38:331–336.

62 Metcalf D, Greenhalgh CJ, Viney E, Willson TA, Starr R, Nicola NA, Hilton DJ, Alexander WS: Gigantism in mice lacking suppressor of cytokine signalling-2. Nature 2000;405:1069–1073.

63 Greenhalgh CJ, Rico-Bautista E, Lorentzon M, Thaus AL, Morgan PO, Willson TA, Zervoudakis P, Metcalf D, Street I, Nicola NA, Nash AD, Fabri LJ, Norstedt G, Ohlsson C, Flores-Morales A, Alexander WS, Hilton DJ: SOCS2 negatively regulates growth hormone action in vitro and in vivo. J Clin Invest 2005;115:397–406.

64 Vajo Z, Francomano CA, Wilkin DJ: The molecular and genetic basis of fibroblast growth factor receptor-3 disorders: the achondroplasia family of skeletal dysplasias, Muenke craniosynostosis, and Crouzon syndrome with acanthosis nigricans. Endocr Rev 2000;21:23–39.

65 Olney RC, Bukulmez H, Bartels CF, Prickett TC, Espiner EA, Potter LR, Warman ML: Heterozygous mutations in natriuretic peptide receptor-B (NPR2) are associated with short stature. J Clin Endocrinol Metab 2006;91:1229–1232.

66 Tanner JM: The regulation of human growth. Child Dev 1963;34:817–847.

67 Cutler GB Jr: The role of estrogen in bone growth and maturation during childhood and adolescence. J Steroid Biochem Mol Biol 1997;61:141–144.

68 Nilsson O, Baron J: Impact of growth plate senescence on catch-up growth and epiphyseal fusion. Pediatr Nephrol 2005;20:319–322.

69 Nilsson O, Marino R, De Luca F, Phillip M, Baron J: Endocrine regulation of the growth plate. Horm Res 2005;64:157–165.

70 Emons JA, Boersma B, Baron J, Wit JM: Catch-up growth: testing the hypothesis of delayed growth plate senescence in humans. J Pediatr 2005;147:843–846.

71 Davenport ML, Crowe BJ, Travers SH, Rubin K, Ross JL, Fechner PY, Gunther DF, Liu C, Geffner ME, Thrailkill K, Huseman C, Zagar AJ, Quigley CA: Growth hormone treatment of early growth failure in toddlers with Turner syndrome: a randomized, controlled, multicenter trial. J Clin Endocrinol Metab 2007;92:3406–3416.

72 Quigley CA: Growth hormone treatment of non-growth hormone-deficient growth disorders. Endocrinol Metab Clin North Am 2007;36:131–186.

73 Rao E, Weiss B, Fukami M, Rump A, Niesler B, Mertz A, Muroya K, Binder G, Kirsch S, Winkelmann M, Nordsiek G, Heinrich U, Breuning MH, Ranke MB, Rosenthal A, Ogata T, Rappold GA: Pseudoautosomal deletions encompassing a novel homeobox gene cause growth failure in idiopathic short stature and Turner syndrome. Nat Genet 1997;16:54–63.

74 Huber C, Rosilio M, Munnich A, Cormier-Daire V: High incidence of SHOX anomalies in individuals with short stature. J Med Genet 2006; 43:735–739.

75 Blum WF, Crowe BJ, Quigley CA, Jung H, Cao D, Ross JL, Braun L, Rappold G: Growth hormone is effective in treatment of short stature associated with short stature homeobox-containing gene deficiency: two-year results of a randomized, controlled, multicenter trial. J Clin Endocrinol Metab 2007;92:219–228.

76 Rappold G, Blum WF, Shavrikova EP, Crowe BJ, Roeth R, Quigley CA, Ross JL, Niesler B: Genotypes and phenotypes in children with short stature: clinical indicators of SHOX haploinsufficiency. J Med Genet 2007;44:306–313.

77 Marchini A, Rappold G, Schneider KU: SHOX at a glance: From gene to protein. Arch Physiol Biochem 2007;113:116–123.

78 Ogata T, Matsuo N, Nishimura G: SHOX haploinsufficiency and overdosage: impact of gonadal function status. J Med Genet 2001;38:1–6.

Primus-E. Mullis
Paediatric Endocrinology, Diabetology and Metabolism
University Children's Hospital, Inselspital
CH–3010 Bern (Switzerland)
Tel. +41 31 632 9552, Fax +41 31 632 9550, E-Mail primus.mullis@insel.ch

Mullis

Hindmarsh PC (ed): Current Indications for Growth Hormone Therapy, ed 2, revised.
Endocr Dev. Basel, Karger, 2010, vol 18, pp 83–91

Clinical Considerations in Using Growth Hormone Therapy in Growth Hormone Deficiency

Michael B. Ranke

University Children's Hospital, Paediatric Endocrinology and Diabetes, Tübingen, Germany

Abstract

Growth hormone (GH) replacement therapy has been the standard medical practice for 50 years in treating GH-deficient patients. When recombinant human GH became available around the mid-1980s, daily GH injections via the subcutaneous route started. The GH dosage corresponds to the daily spontaneous secretion (e.g. 20–30 µg/kg body weight in the prepubertal stage; 2–4 times during puberty). Since the growth process is long, the effects of therapeutic strategies can only be observed many years later. Reports of long-term results show that, over the decades, there is a trend towards diagnosing/treating younger patients and that they are less GH-deficient and not as short as patients in the past. Today about 80–90% of these patients attain normal height. We have proposed that the customary 'fixed-dose wait-and-see' approach should be relinquished in favor of an adaptive strategy of dosing which focuses on the individual responsiveness of each patient. The prediction models derived from observations of large cohorts of patients have become essential in determining the responsiveness. This approach not only optimizes the outcomes but also the cost of treatment. Further efforts are needed to incorporate the existing evidence into the diagnostic procedure and treatment of GH deficiency.

Introduction

When human growth hormone (GH) extracted from pituitaries became available in the late 1950s, Dr. Maurice Raben (1958) [1] proved that growth promotion was possible by injecting a GH-deficient patient with GH. This was a significant feat, particularly in light of the fact that GH was difficult to measure by means of the bioassays available before 1963, since they were tedious and imprecise. Advancements during the 50 years that followed Raben's landmark trial have led to a multitude of discoveries and deepened our understanding of the complex regulation of GH and its comprehensive biological functions throughout the human lifespan. One fundamental discovery was that, at the level of the growth plate, the growth-promoting effect of GH was mediated directly by GH as well as indirectly by the insulin-like growth factor-1 (IGF-1), a GH-dependent hormone. New imaging techniques (e.g. CT, MRI) and molecular genetic methods have further widened the scope of investigations into the various causes of an impairment in GH secretion. Recombinant (22 kDa) human GH, available since 1987, has become the standard replacement therapy in GH-deficient patients.

To date, however, some uncertainty regarding the optimization of the diagnostic procedure and the most effective use of GH, in terms of efficacy, safety and costs, remain. The aim of this chapter is to discuss the most importance aspects involved in the treatment of growth hormone deficiency (GHD) in children and adolescents.

Growth Hormone Therapy

A few reports in the literature confirm that untreated GH-deficient patients have severe short stature [2–5], with spontaneous adult height in the order of –4.5 SDS. However, no documentation is available on the natural history of the various components of statural growth – height, weight, proportions, head size – in children with isolated congenital GHD that would have allowed insights to be gained into the specific effect of GH during all phases of childhood development.

GH Preparations
In the first decade following Raben's proof-of-concept study, attempts were made to increase the yield of GH extracted from a pituitary and to improve the purity of the product by combining several separation techniques. National agencies were established in the USA, UK and France for the collection of pituitaries and purification of GH. In addition, the commercial production of pituitary GH had also begun. The potency of the preparations was approximately (1–)2 IU/mg. The production of pithGH was brought to an end in 1985, after some recipients developed Creutzfeldt-Jakob disease. Several companies started marketing recombinant human GH from 1987 onwards, with the products comprising 22 kDa hGH with a potency of 3 IU/mg.

Application and Dose of GH
GH has traditionally been applied via the intramuscular route, on the assumption that complete absorption would thus be ensured. It was also suspected that a higher frequency of antibodies to GH would occur through subcutaneous injection [6]. Initially, GH was given once or twice per week. However, it was then shown that

more frequent injections led to a higher gain in height during the first year of treatment [7] and Kastrup et al. [8] demonstrated that daily doses of GH, injected through the subcutaneous route, was the most suitable mode of application. The procedure is now practically painless due to modern injection devices and highly sophisticated needles. Since GH secretion mainly occurs during the night, it has become common practice to carry out the GH injections in the evening.

At first, the GH dosages for GH-deficient patients were based on rather crude estimates. It was only after multiple sampling techniques and deconvolution analysis were developed that it became possible to determine the true secretion rates of GH in prepubertal and pubertal children [9, 10]: the daily secretion rate during the prepubertal age is about 0.02 (0.06 IU)/kg body weight; during early puberty this amount rises by a factor of 2–4 due to sex steroids (estrogens). Since GH is secreted in a pulsatile manner, it is not appropriate to compare the area under the curve (AUC) for endogenous secretion in normal individuals with that of patients who inject GH. This is probably one of the reasons why long-acting forms of GH have not yet found their way into the market.

The dose dependence of growth, during the first year of treatment, was first reported by Preece et al. [11]. The authors had studied a group of patients with mixed causes of GHD, and found that an increase in the weekly GH dose, from 10 to 20 IU, led to an increase in height velocity by a factor of 1.3. These findings confirmed that the relationship between dose and response was not linear. Frasier et al. [12] went a step further by showing a positive relationship between the growth response (cm/years) and the logarithms of GH dosages between 30 and 200 mIU/kg. Dosing per kg body weight can be expressed in different ways: μg/kg day (/1,000) = mg/kg day (×7) = mg/kg week (×3) = IU/kg week. Some clinicians prescribe GH per m^2 body surface, and 1 m^2 in a child equals a weight of roughly 30 kg.

Results of GH Replacement
It is important that the GH treatment of GH-deficient children is guided by the following key objectives: (1) normal height should be reached as soon as possible, and be maintained during the growth period, (2) puberty should start at a normal timing, and pubertal growth and development should be normal, (3) the adult height attained should be within the normal range, (4) the risks of therapy should be minimized, and (5) the aims of treatment should be achieved at the minimum cost.

Until recently, investigators tended to embark on a particular treatment strategy (e.g., GH dose) and followed the treatment of their patients up to adult height. Several reports about the outcomes in large series of patients [5, 13–17] are available, along with review articles [18–21]. A few representative results of this approach are listed in table 1. The data show the following characteristics

Table 1. Height development in children with GHD treated to adult height. Representative results from the past thre
decades

Author (year)	Group (M/F) Puberty (spontaneous/ induced)	n	Dosage IU/week	Age start	Start Ht (SDS)	Adult Ht (SDS)	Gain H (SDS)
Burns, 1981 [14]	M (spontan.)	30	15–20	13.8	–4.4	–2.1	2.3
	F (spontan.)	9		11.7	–5.0	–2.9	2.1
	M (induced)	11		13.2	–5.1	–1.4	3.7
	F (induced)	5		12.4	–5.5	–1.4	4.1
Lenko, 1982 [15]	M (spontan.)	5	8–12	12.7	–4.0	–2.4	1.6
	F (spontan.)	5		13.2	–6.0	–4.7	1.3
	M (induced)	7		14.6	–4.8	–1.2	3.6
	F (induced)	7		15.9	–4.3	–2.2	2.2
Wit, 1996 [5]	M (spontan.)	53	8	12.3	–4.0	–2.3	1.7
	F (spontan.)	21		9.2	–4.2	–2.6	1.6
	M (induced)	41		13.1	–3.7	–1.8	1.9
	F (induced)	21		13.3	–3.6	–1.8	1.8
Ranke, 1997 [17]	M (spontan.)	66	0.57	10.5	–2.7	–1.3	1.4
	F (spontan.)	51	0.58	9.9	–2.8	–0.5	2.3
	M (induced)	64	0.46	9.9	–2.9	–1.2	1.7
	F (induced)	14	0.64	6.8	–2.7	–0.9	1.8
Tanaka, 1999 [21]	M (isolated)	76	0.5	12.2	–2.8	–1.4	1.5
	F (isolated)	52		11.6	–3.0	–1.3	1.7
	M (MPHD)	11		9.5	–2.8	–0.1	2.5
	F (MPHD)	13		12.5	–2.8	–1.1	1.5
Reiter, 2006 [16]	M (isolated)	351	0.66	10.1	–2.4	–0.8	1.6
	F (isolated)	200	0.60	9.3	–2.6	–1.0	1.6
	M (MPHD)	257	0.54	7.2	–3.4	–0.7	2.7
	F (MPHD)	172	0.54	8.0	–2.9	–1.1	1.8
Westphal, 2008 [13]	M (spontan.)	294	0.69	9.1	–2.7	–0.9	1.8
	F (spontan.)	107	0.69	8.0	–2.9	–0.8	2.1

and changes over time: (1) Patients treated at the time pituitary GH was first
used (years before 1985) had the lowest height recorded at treatment start. There
is a trend to suggest that GH-deficient patients are now not as short as patients
in the past. (2) The ratio between male and female patients is about 1.5:1, with
girls being younger and shorter at the start of GH treatment. (3) The GH dosage
has been increasing in general. (4) More children reach a normal height, today;

however, the overall gain in height through GH treatment is lower. (5) Children with combined pituitary deficits and/or induced puberty tend to have a better height outcome.

It must be noted, however, that even today the mean adult height of patients treated for many years is approximately 1 SD below the mean of the reference population. In other words, a considerable number of patients (15–20%) remain short.

Prediction Models for the Treatment of GHD
In view of the above, it is imperative that a new approach is taken, in which factors influencing the response to GH are analyzed. The response (the observed growth) during the first year of GH treatment has been investigated in several small studies [22–25] and a number of factors were identified which were associated both with the characteristics of the treated child, i.e. anthropometry (age, fat mass, weight, bone age) and biochemistry (levels of GH, IGF-1); as well as to the modalities of treatment (GH dose, frequency of injections). Obviously, any evaluation of the magnitude of the *response* must be seen within the context of the specific characteristics of the individual. The ability of an individual to respond (e.g., in terms of height gain) to a specific modality of treatment is the *'responsiveness'*. The closest probable response to treatment in a GH-deficient patient can be illustrated through a prediction model. Such models are, in fact, mathematical algorithms (regression equation) which describe the most likely response (mean and its variation) during a given phase of therapy, for instance, the gain in height (cm/year) during the first year of prepubertal therapy. Such algorithms were devised on the basis of large databases comprising a range of information (potential predictors) relating to the GH-deficient patient at the time treatment was being carried out. The selection of predictors and the development of mathematical algorithms is a complex but well-established process [26]. The author's group has developed prediction models for each prepubertal year (first to eighth) of GH treatment [27]. A prediction model is also available for the total pubertal growth phase [28]. The analyses were based on large numbers of patients and explain 30–70% of the responses in the relevant years. The numerical components of the algorithms are listed in table 2 and can be used to calculate the height velocity (cm/year) [height (cm) for total pubertal growth, respectively] for the respective years/time period.

Here we give an example for growth during the first year of therapy:

Height velocity 1st year (model A): cm/year
= 14.6 + (age [years] × –0.32) + (max GH [ln µg/l] × –1.37) +
(birth weight [SDS] × 0.32) + (GH dose [ln IU/kg week] × 1.62) +
(Ht – MPH [SDS] × –0.4) + (weight [SDS] × 0.29).

Table 2. Components of the KIGS algorithms for height velocity (HV) in GH-treated children with idiopathic GHD

Estimates	Dimension	Prepubertal year					Total pubertal growth
		1 (model A)	1 (model B)	2	3	4–8	
Parameter							
Intercept	–	14.6	12.4	5.7	5.6	6.0	48.3
Age at GH start	years	–0.32	–0.36	–0.09	–0.10	–0.05	–
Maximum GH level (tests)	ln µg/l	–1.37	–	–	–	–	–
Birth weight	SDS	0.32	0.47	–	–	–	–
GH dose	ln IU/kg/week	1.62	1.54	0.63	0.66	0.87	
Height–MPH	SDS	–0.40	–0.60	–	–	–	–1.32
Weight	SDS	0.29	0.28	0.24	0.30	0.40	–
HV (previous year)	cm/year	–	–	0.31	0.32	0.21	–
Age at puberty	years	–	–	–	–	–	–3.02
GH dose	IU/kg/week	–	–	–	–	–	6.46
Sex	boy = 1	–	–	–	–	–	11.3
Statistics							
n		592	592	572	333	179	303
R^2		0.61	0.45	0.40	0.37	0.30	0.70
Error SD	cm/year	1.46	1.72	1.19	1.05	0.95	–
	cm	–	–	–	–	–	4.19

The models indicate a number of important components during the first prepubertal year of growth, in particular the degree of GHD (the more severe, the better the response), age (the younger, the better the response), and the distance of height to mid-parent height [ht – MPH] (the shorter in comparison to target height, the better the response). The GH dosage plays a relatively minor role. In the subsequent prepubertal years, the most important predictor is the response during the preceding year (once a good response, always a good response), followed by the GH dose. Components of total pubertal growth include age (the younger, the higher the pubertal growth) and height at puberty onset (the shorter, the greater the pubertal growth). Thus a child who is younger and shorter at puberty onset will – as in normal growth – gain a greater fraction of the growth-to-target height, and vice versa. The prediction model can also be used to retrospectively confirm the observations from studies [29, 30], namely that there is only little gain in height if the GH dose is increased. By calculating the 'index of responsiveness' (IR) = [(observed HV(or Ht) – predicted HV(or Ht))/Error SD] it is possible to observe the extent of the deviation from the normal growth of the reference cohort.

These models can be used in many ways. First, in counseling those directly involved (patients, parents, family doctors), the models can be used to inform the family about what can reasonably be expected from treatment, thus contributing to better concordance. Second, if there are significant negative deviations than expected (IR <–1.0 SD) one should examine other reasons, such as wrong diagnosis, non-concordance, other hormonal deficits, additional disorders, technical problems. Thus an increase in the dosage and unnecessary costs can be avoided in patients with poor responsiveness. Conversely, lower amounts of GH can be applied in patients who respond well. Third, therapy (dosing) can be adjusted as required. A patient with a lower growth potential, e.g. due to a delay in the diagnosis, could optimally receive a higher dose for the remaining duration of treatment.

Conclusions

The diagnosis of GHD must be based on the best methodology and empirical evidence available. This would involve a degree of methodological standardization that is higher than the existing level. It is only through such an approach that it is possible to treat patients appropriately and to avoid unnecessary therapeutic interventions. After the diagnosis is established, GH replacement needs to be optimized, depending on the individual requirements. Prediction models are a tool towards achieving this aim. The models may need further refinement, i.e. the inclusion of further anthropometrical, biochemical and molecular genetic parameters. It is striking how parameters that can be accurately recorded in clinical practice, can be transformed into a powerful clinical tool. Clearly, the 'wait-and-see' attitude in treating GHD is no longer appropriate. Treating physicians should evaluate the response – and responsiveness – of the patient and then adapt treatment accordingly. This will hopefully lead to a normalization of height and other components which are impaired by GHD, so that at the close of adolescent life a successful transition to adult care can take place.

References

1 Raben MS: Treatment of a pituitary dwarf with human growth hormone. J Clin Endocrinol Metab 1958;18:901–903.

2 Van der Werff ten Bosch JJ, Bot A: Growth of males with idiopathic hypopituitarism without growth hormone treatment. Clin Endocrinol (Oxf) 1990;32:707–717.

3 Ranke MB: A note on adults with growth hormone deficiency. Acta Paediatr Scand Suppl 1987; 331:80–82.

4 Rimoin DL, Merimee TJ, Rabinowitz D, McKusick VA: Genetic aspects of clinical endocrinology. Recent Prog Horm Res 1968;24:365–437.

5 Wit JM, Kamp GA, Rikken B: Spontaneous growth and response to growth hormone treatment in children with growth hormone deficiency and idiopathic short stature. Pediatr Res 1996; 39:295–302.

6 Underwood LE, Voina SJ, Van Wyk JJ: Restoration of growth by human growth hormone (Roos) in hypopituitary dwarfs immunized by other human growth hormone preparations: clinical and immunological studies. J Clin Endocrinol Metab 1974;38:288–297.

7 Milner RD, Russell-Fraser T, Brook CG, Cotes PM, Farquhar JW, Parkin JM, Preece MA, Snodgrass GJ, Mason AS, Tanner JM, Vince FP: Experience with human growth hormone in Great Britain: the report of the MRC Working Party. Clin Endocrinol (Oxf) 1979;11:15–38.

8 Kastrup KW, Christiansen JS, Andersen JK, Orskov H: Increased growth rate following transfer to daily subcutaneous administration from three weekly intramuscular injections of hGH in growth hormone-deficient children. Acta Endocrinol (Copenh) 1983;104:148–152.

9 Albertsson-Wikland K, Rosberg S, Libre E, Lundberg LO, Groth T: Growth hormone secretory rates in children as estimated by deconvolution analysis of 24-hour plasma concentration profiles. Am J Physiol 1989;257:E809–E814.

10 Martha PM Jr, Gorman KM, Blizzard RM, Rogol AD, Veldhuis JD: Endogenous growth hormone secretion and clearance rates in normal boys, as determined by deconvolution analysis: relationship to age, pubertal status, and body mass. J Clin Endocrinol Metab 1992;74:336–344.

11 Preece MA, Tanner JM, Whitehouse RH, Cameron N: Dose dependence of growth response to human growth hormone in growth hormone deficiency. J Clin Endocrinol Metab 1976;42:477–483.

12 Frasier SD, Costin G, Lippe BM, Aceto T Jr, Bunger PF: A dose-response curve for human growth hormone. J Clin Endocrinol Metab 1981; 53:1213–1217.

13 Westphal O, Lindberg A: Swedish KIGS National Board. Acta Paediatr 2008;97:1698–1706.

14 Burns EC, Tanner JM, Preece MA, Cameron N: Final height and pubertal development in 55 children with idiopathic growth hormone deficiency, treated for between 2 and 15 years with human growth hormone. Eur J Pediatr 1981;137:155–164.

15 Lenko HL, Leisti S, Perheentupa J: The efficacy of growth hormone in different types of growth failure. An analysis of 101 cases. Eur J Pediatr 1982; 138:241–249.

16 Reiter EO, Price DA, Wilton P, Albertsson-Wikland K, Ranke MB: Effect of growth hormone (GH) treatment on the near-final height of 1,258 patients with idiopathic GH deficiency: analysis of a large international database. J Clin Endocrinol Metab 2006;91:2047–2054.

17 Ranke MB, Price DA, Albertsson-Wikland K, Maes M, Lindberg A: Factors determining pubertal growth and final height in growth hormone treatment of idiopathic growth hormone deficiency. Analysis of 195 patients of the Kabi Pharmacia International Growth Study. Horm Res 1997;48:62–71.

18 Growth Hormone Research Society: Consensus guidelines for the diagnosis and treatment of growth hormone (GH) deficiency in childhood and adolescence: summary statement of the GH Research Society. GH Research Society. J Clin Endocrinol Metab 2000;85:3990–3993.

19 American Association of Clinical Endocrinologists Growth Hormone Task Force: American Association of Clinical Endocrinologists medical guidelines for clinical practice for growth hormone use in adults and children – 2003 update. Endocr Pract 2008;9:64–76.

20 Guyda HJ: Four decades of growth hormone therapy for short children: what have we achieved? J Clin Endocrinol Metab 1999;84:4307–4316.

21 Tanaka T, Cohen P, Clayton PE, Laron Z, Hintz RL, Sizonenko PC: Diagnosis and management of growth hormone deficiency in childhood and adolescence. 2. Growth hormone treatment in growth hormone deficient children. Growth Horm IGF Res 2002;12:323–341.

22 Albertsson-Wikland K: The effect of human growth hormone injection frequency on linear growth rate. Acta Paediatr Scand Suppl 1987;337: 110–116.

23 Wit JM, van't Hof MA, Van den Brande JL: The effect of human growth hormone therapy on skinfold thickness in growth hormone-deficient children. Eur J Pediatr 1988;147:588–592.

24 Sherman BM, Frane J, Johanson AJ, Kaplan SL: Predictors of response to treatment with methionyl human growth hormone; in Underwood L (ed): Human Growth Hormone: Progress and Challenges. New York, Dekker, 1988, pp 131–144.

25 Rolland A, Mugnier E, Rappaport R, Job JC: Le traitement du nanisme hypophysaire par l'hormone de croissance humaine. Arch Fr Pédiatr 1980,37: 659–665.

26 Ranke MB, Lindberg A: Predicting growth in response to growth hormone treatment. Growth Horm IGF Res 2009;19:1–11.

27 Ranke MB, Lindberg A, Chatelain P, Wilton P, Cutfield W, Albertsson-Wikland K, Price DA: Derivation and validation of a mathematical model for predicting the response to exogenous recombinant human growth hormone (GH) in prepubertal children with idiopathic GH deficiency. KIGS International Board. Kabi Pharmacia International Growth Study. J Clin Endocrinol Metab 1999;84: 1174–1183.

28 Ranke MB, Lindberg A, Martin DD, Bakker B, Wilton P, Albertsson-Wikland K, Cowell CT, Price DA, Reiter EO: The mathematical model for total pubertal growth in idiopathic growth hormone (GH) deficiency suggests a moderate role of GH dose. J Clin Endocrinol Metab 2003;88: 4748–4753.

29 Coelho R, Brook CG, Preece MA, Stanhope RG, Dattani MT, Hindmarsh PC: A randomised study of two doses of biosynthetic human growth hormone on final height of pubertal children with growth hormone deficiency. Horm Res 2008;70: 85–88.

30 Mauras N, Attie KM, Reiter EO, Saenger P, Baptista J: High dose recombinant human growth hormone (GH) treatment of GH-deficient patients in puberty increases near-final height: a randomized, multicenter trial. Genentech Inc Cooperative Study Group. J Clin Endocrinol Metab 2000;85: 3653–3660.

Michael B. Ranke, MD
Paediatric Endocrinology and Diabetology, University Children's Hospital
Hoppe-Seyler-Strasse 1, DE–72076 Tübingen (Germany)
Tel. +49 7071 2983417, Fax +49 7071 294157
E-Mail michael.ranke@med.uni-tuebingen.de

Hindmarsh PC (ed): Current Indications for Growth Hormone Therapy, ed 2, revised.
Endocr Dev. Basel, Karger, 2010, vol 18, pp 92–108

Current Indications for Growth Hormone Therapy for Children and Adolescents

Erick Richmond[a] · Alan D. Rogol[b]

[a]Pediatric Endocrinology, National Children's Hospital, San José, Costa Rica; [b]Pediatric Endocrinology,
Riley Hospital, Indiana University School of Medicine, Indianapolis, Ind., and University of Virginia,
Charlottesville, Va., USA

Abstract

Growth hormone (GH) therapy has been appropriate for severely GH-deficient children and adolescents since the 1960s. Use for other conditions for which short stature was a component could not be seriously considered because of the small supply of human pituitary-derived hormone. That state changed remarkably in the mid-1980s because of Creutzfeldt-Jakob disease associated with human pituitary tissue-derived hGH and the development of a (nearly) unlimited supply of recombinant, 22 kDa (r)hGH. The latter permitted all GH-deficient children to have access to treatment and one could design trials using rhGH to increase adult height in infants, children and adolescents with causes of short stature other than GH deficiency, as well as trials in adult GH-deficient men and women. Approved indications (US Food and Drug Administration) include: GH deficiency, chronic kidney disease, Turner syndrome, small-for-gestational age with failure to catch up to the normal height percentiles, Prader-Willi syndrome, idiopathic short stature, SHOX gene haploinsufficiency and Noonan syndrome (current to October 2008). The most common efficacy outcome in children is an increase in height velocity, although rhGH may prevent hypoglycemia in some infants with congenital hypopituitarism and increase the lean/fat ratio in most children – especially those with severe GH deficiency or Prader-Willi syndrome. Doses for adults, which affect body composition and health-related quality of life, are much lower than those for children, per kilogram of lean body mass. The safety profile is quite favorable with a small, but significant, incidence of raised intracranial pressure, scoliosis, muscle and joint discomfort, including slipped capital femoral epiphysis. The approval of rhGH therapy for short, non-GH-deficient children has validated the notion of GH sensitivity, which gives the opportunity to some children with significant short stature, but with normal stimulated GH test results, to benefit from rhGH therapy and perhaps attain an adult height within the normal range and appropriate for their mid-parental target height (genetic potential).

Copyright © 2010 S. Karger AG, Basel

Introduction

Human growth hormone (hGH) is a single-chain, 191-amino-acid protein of 22 kDa molecular weight. It is synthesized, stored and released from the somatotropes of the anterior pituitary gland. Multiple additional molecular variants, including a 20-kDa form produced by the gene deletion of 14 amino acids and other post-translational isoforms, for example glycosylated and sulfated forms, of unknown physiological significance exist within the systemic circulation [1].

GH is relatively species-specific since only primate GH has efficacy in the human [2, 3]. None of the small trials with animal GH or enzymatically-produced fragments from animal GH showed growth-promoting efficacy in the human [4].

hGH was first extracted in the late 1940s [5] and administered to humans beginning in 1958 [6, 7]. Treatment of both children and adults was reported in that first clinical trial, although the latter took a number of years for confirmation of efficacy since so little hormone was available (essentially one pituitary per child per day). Thus, in the early years only profoundly GH-deficient children received treatment. Much of the time so little hormone was available that many children were treated 6 or 9 months a year permitting other, equally needy, children to reap some benefit from hGH therapy.

That all changed in the early 1980s when it became clear that Creutzfeldt-Jakob disease could be transmitted by human brain tissue and recombinant (r)hGH became available [8]. At that time a seemingly endless supply of the pure 22-kDa hormone permitted all GH-deficient children to have access to treatment and one could now design trials using rhGH to increase adult height in infants, children and adolescents with causes of short stature other than GH deficiency, as well as trials in adult GH-deficient men and women. The next 20 years led to at least 7 additional indications in children and 3 new ones in adults (table 1).

hGH is administered to promote linear growth in short children. The following are the US Food and Drug Administration (FDA)-approved indications for GH (current to October 2008) and in parentheses the indications approved in Europe (E) by the European Agency for the Evaluation of Medicinal Products (EMEA).

The most common efficacy outcome in infants, children and adolescents is an increase in linear growth, although rhGH may prevent hypoglycemia in some infants with congenital hypopituitarism and increase the lean/fat ratio in children with the Prader-Willi syndrome (see below).

The dose-response curve for height gain versus dose of rhGH (log scale) is rather flat in those with GH deficiency [9], even through the much higher doses (including 100 µg/kg/day) administered more recently [10]. The dose for adults which affects body composition and health-related quality of life are much lower per kilogram of body mass.

Children
Growth hormone deficiency (E)
Chronic kidney disease (E)
Turner syndrome (E)
Small-for-gestational age infants who fail to catch up to the normal growth percentiles (E)
Prader-Willi syndrome (E)
Idiopathic short stature
SHOX gene haploinsufficiency (E)
Noonan syndrome
Adults
Growth hormone deficiency (E)
HIV/AIDS wasting
Short bowel syndrome

In this review, the studies that were selected for analysis were mainly randomized controlled studies, with the larger number of patients and the longest treatment duration of the available publications.

Growth Hormone Deficiency (GHD)

The administration of GH to treat children with sort stature resulting from GHD or GH insufficiency has now accrued over 40 years of clinical experience. In 1985, the US FDA approved the use of rhGH for the treatment of GHD and remains the primary indication for GH treatment in childhood. GHD is fundamentally a clinical diagnosis, based upon auxologic features. Assessment of laboratory tests, whether static, for example the measurement of IGF-1 and IGFBP-3, or dynamic, for example, secretagogue-stimulated GH secretion is confirmatory [11]. Radiologic evaluation of the hypothalamus and pituitary (CT scan or MRI) is helpful in patients with suspected congenital GHD or to detect space-occupying lesions (see Chapters 1 and 3 for fuller discussion of these points).

The primary objectives of therapy of GHD are normalization of height during childhood and attainment of adult height within the normal range. The greater efficacy demonstrated in recent large international databases [12, 13] compared with the initial studies of GH treatment in GHD patients (table 2) probably reflects the combined effects of the higher GH dose, the more physiologic injection frequency, and the younger age at initiation of treatment. Current consensus guidelines recommend a dose in the range of 0.025–0.05 mg/kg/day [21, 22]. Under

Patients n	Mean age at start, years	Mean height SD at start	GH dose mg/kg/day	Mean duration of treatment years	Estimated height gain SD	Ref.
88	8.2	–2.9	0.023–0.030*	9.9	2.4	14
121	10.8	–3.2	0.043*	7.8	2.5	15
1,034	12.2	–2.8	0.028–0.043*	4.6	1.3	16
932	12.0	–2.8	NA	4.9	1.0	17
25	10.0	–4.5	0.020–0.028*	8.6	1.1	18
18	12.0	–5.6	0.020–0.023	6.2	2.8	19
13	3.6	–4.1	0.023–0.031*	12.2	3.2	20

NA = Not available.
* GH dose was administered 3–7 times per week.

special circumstances, higher doses may be required, including adolescents with late diagnosis and diminished period of time for catch-up growth [23]. Recently it has been proposed that IGF-1-based GH dosing may improve growth responses, although at higher average GH doses [24].

Significant side effects of GH treatment in children are uncommon. These include idiopathic intracranial hypertension, prepubertal gynecomastia, arthralgia, and edema [21, 22]. Management of these side effects may include either transient reduction of dosage or temporary discontinuation of GH. In the absence of other risk factors, there is no evidence that the risk of leukemia, brain tumor recurrence, slipped capital femoral epiphysis or diabetes are increased in recipients of long-term GH treatment.

GHD may or not persist into adulthood. After the attainment of adult height, retesting of the GH-IGF-1 axis is recommended prior to the decision to continue or not with GH treatment through adulthood [25]. GH has major metabolic actions, which are important for body composition and health in adults as well as in children.

Chronic Renal Insufficiency (CRI), also Known as Chronic Kidney Disease (CKD)

Growth failure is common in children with CRI, which was the first non-GHD growth disorder for which GH was approved for use by the US FDA in 1993 and by the EMEA in 1995.

Malnutrition and alterations in the GH-IGF axis seem to be the most frequent and most important factors contributing to the degree of growth disturbance.

Table 3. GH trials in children with CRI

Patients n	Mean age at start, years	Mean height SD at start	GH dose mg/kg/day	Mean duration of treatment years	Estimated height gain SD	Ref.
38	10.4	–3.1	0.047	5.3	1.4	31
183	8.2	–2.7	0.027–0.054	5.3	0.4	32
45	7.8	–3.0	0.050	8.0	2.6	33
30	5.6	–2.7	0.050	1.0	0.8	34
55	6.0	–2.9	0.050	2.0	1.4	35

* GH dose was administered 3–7 times per week.

Other factors are: the specific etiology of the renal disease, acid-base disturbances, secondary hyperparathyroidism, age, duration of renal disease and treatment modalities [26]. Long-term treatment with high-dose glucocorticoids in children leads to growth failure and protein catabolism and glucocorticoids interfere with the integrity of the GH-IGF axis at various levels [27].

Children with CRI who do not receive GH treatment attain an adult height below the expected genetic potential in the majority and below –2 SDS in about half of affected individuals [28]. Approval of GH treatment in patients with growth failure associated with CRI was based on only short-term data, but subsequent studies showed that GH has some short-term effect although adult height are perhaps less impressive [29, 30–35] (table 3). GH treatment is approved only for CRI before renal transplantation, but a number of studies have successfully treated children after transplantation [36, 37].

Although GH treatment in patients with CRI is considered safe [38], some studies have raised the possibility of deterioration of renal function, especially in those who have a previous history of renal graft rejection [39], and alterations in glucose metabolism [40]. Careful monitoring is recommended, especially in children who have a family history of diabetes or those receiving concomitant glucocorticoid therapy.

Turner Syndrome (TS)

TS occurs in approximately 1 in every 2,500 live-born females, making it one of the most common chromosomal disorders [41]. The diagnosis of TS requires the presence of characteristic physical findings [42, 43] in addition to a complete or partial absence of the second sex chromosome, with or without cell line mosaicism [44]. Untreated adult women with TS achieve an average adult height 20 cm shorter than their mid-parental (target) height [45, 46].

Table 4. GH trials in girls with TS

Patients n	Mean age at start, years	Mean height SD at start	GH dose mg/kg/day	Mean duration of treatment years	Estimated height gain SD	Ref.
70	9.3	−3.2	0.054*	7.6	1.1	49
60	9.5	−3.0	0.050	5.9	1.0	50
232	9.7	−3.1	0.039–0.051	5.5	1.1	51
154	10.3	−3.2	0.050	5.7	1.1	52
14	10.6	−3.2	0.037–0.111	4.0	1.1	53
68	6.6	−2.8	0.045–0.090	8.6	1.7	54
188	11.7	−3.7	0.025–0.056	2.4	0.9	55
704	11.9	−3.4	0.043	5.0	1.2	56
88	2.0	−1.4	0.050	2.0	1.1	57
60	10.9	−2.7	0.047	6.8	0.9	58

* GH dose was 0.125 mg/kg three times per week.

Despite some evidence for subtle alterations in IGF-1 physiology, the growth deficit in individuals with TS is believed to result mainly from haploinsufficiency of one copy of the SHOX gene, located within the pseudoautosomal region on the distal short arm of the X (and Y) chromosome [47], and discussed in further detail in the section on SHOX deficiency (see below).

Despite dozens of publications of GH treatment in TS patients over the last two decades, a recent Cochrane Center review identified only four studies in which GH treatment was compared in a randomized fashion with a concurrent non-treatment or placebo control [48]. In these studies, patients treated to adult or near-adult height achieved average height gains ranging from about 5 to 8 cm over periods of treatment ranging from 5.5 to 7.6 years, and the doses varied between 0.039 and 0.054 mg/kg/day (table 4) [49–52]. Other studies of GH treatment in TS patients have variable results probably due to differences in methodology with respect to patient selection, the type of controls used, rhGH dose and frequency, concomitant treatment with oxandrolone or estrogen and analytic methods (table 4) [53–58].

Pediatric and adolescent patients with TS appear to be at increased risk for some adverse events associated with GH treatment, including intracranial hypertension, scoliosis, pancreatitis, and slipped capital femoral epiphysis, compared with other populations who receive GH treatment [59]. They are also at risk for events known to be associated with TS, including autoimmune disorders, and likely type 1 diabetes mellitus, as well as hearing loss, hypertension and aortic dissection with rupture [60, 61].

Children Born Small-for-Gestational Age (SGA), Who Fail to Catch Up to the Normal Growth Channels

The definition of SGA varies in the pediatric literature. A recent consensus statement recommended that SGA should be defined as a weight and/or length less than –2 SD for the length of pregnancy [62]. Babies can then be subclassified into SGA for weight, SGA for length, or SGA for both weight and length [63], although the response to rhGH therapy did not differ between these groups treated in a trial of 201 SGA babies in the Netherlands [64]. Approximately 90% of infants born SGA undergo spontaneous catch-up growth to return to their genetic potential by the end of the second year of life; however, most catch up occurs by 9–15 months The preterm infant may take 4 years (rarely more) to achieve a height within the normal range [65], the remaining 10% who do not catch up are eligible for rhGH therapy.

The activity of the GH-IGF-1 at birth or during the first weeks of life is not predictive of later growth, and in most patients the endocrine evaluation of growth is not helpful [66]. Later assessment of the GH-IGF-1 axis in the short SGA child may be required if growth velocity is persistently reduced and/or if signs of GH deficiency or hypopituitarism are present [67].

Treatment with rhGH is indicated only when other causes of short stature, including medications that affect growth, chronic diseases, endocrine disorders or syndromes associated with diminished growth and GH therapy is not recommended, have been ruled out.

Experience with the use of rhGH therapy in SGA children started more than three decades ago [68, 69]. The US FDA approved in 2001 the use of GH for the long-term treatment of growth failure in children born SGA, who fail to catch up to the normal growth channels by age 2 years. The recommendation is to administer a dose of up to 0.07 mg/kg/day. In contrast, the EMEA approved GH in 2003 for the treatment of children born SGA after the age of 4 years at a dose of 0.033 mg/kg/day. The reason for waiting until age 4 years is that there is a small possibility of spontaneous catch-up growth between 2 and 4 years of age, especially in those children born prematurely [70].

There is considerable variation in the height gain attained with rhGH therapy in children born SGA, even after adjusting for differences in parental height, age at start of treatment, and duration of treatment [71]. Data from recent randomized rhGH trials in short children born SGA are presented in table 5 [72–75].

The effect of GH on glucose metabolism in children born SGA is of potential concern. Although there is evidence of a tendency to develop higher fasting plasma insulin concentrations and relative insulin resistance during rhGH treatment [76], that phenomenon typically resolves after discontinuation of rhGH treatment [77] although the patients still remain hyperinsulinemic as they are pre-intervention.

Table 5. Randomized GH trials in short children born SGA

Patients n	Mean age at start, years	Mean height SD at start	GH dose mg/kg/day	Mean duration of treatment years	Estimated height gain SD	Ref.
28	5.4	–3.6	0.033–0.067	10.0	1.2	72
77	10.7	–2.8	0.033	7.0	1.3	73
91	12.7	–3.2	0.067	2.7	0.6	74
54	8.1	–3.0	0.033–0.067	7.8	1.8	75

Table 6. GH trials in children with PWS

Patients n	Mean age at start, years	Mean height SD at start	GH dose mg/kg/day	Mean duration of treatment years	Estimated height gain SD	Ref.
35	9.9	–1.1	0.026–0.043	2	0.8	83
15	6.8	–1.6	0.037	1	1.2	84
22	6.9	–1.6	0.030–0.060	9.2	1.9	85
44	4.5	–2.1	0.026–0.043	2	1.6	86

Adverse events due to rhGH therapy are not more common in this population than in other conditions treated with rhGH [78].

Prader-Willi Syndrome (PWS)

PWS, a genetic disorder first described in 1956, is caused by deletion or lack of expression of a portion of the paternally derived chromosome 15 [79]. Affected children are characterized by obesity, hypotonia, short stature, hypogonadism and behavioral abnormalities [80]. Body composition abnormalities including increased fat mass, decreased lean body mass, and low bone density have been demonstrated in patients with PWS [81]. The estimated incidence of this condition is of 1 in 10,000–12,000 births and is the most common of all the syndromes associated with severe obesity [82].

The efficacy of rhGH to increase linear growth and to improve body composition measurements in genetically confirmed patients with PWS irrespective of their GH status is well documented [83–86] (table 6). In 2000 the US FDA approved the use of rhGH for pediatric patients who have growth failure due to PWS and European labeling states 'for improvement of growth and body composition'.

Sudden death has been reported in 17 children with PWS who were treated with rhGH. These children had underlying morbid obesity and respiratory, or possible sleep disorders [87]. Cause and effect remain to be established. Nonetheless, in patients with PWS, correction of underlying airway obstruction and sleep studies should be considered before initiation of rhGH therapy [88]. Because weight gain is a major problem in PWS and the risk of type 2 diabetes is increased, careful monitoring of glucose metabolism is mandatory when rhGH is prescribed [89].

Idiopathic Short Stature (ISS)

ISS or non-GH-deficient short stature, perhaps the most controversial of all the conditions in which rhGH therapy is used, has been approved in the USA. ISS refers to a heterogeneous group of children whose short stature cannot be explained by a defined pathologic process. The diagnosis is based on a process of exclusion, because patients with ISS have no distinguishing clinical or phenotypic features. Most children have a height that is only slightly below the normal range, but others have growth failure as significant as that of GH deficiency [90]. As molecular diagnostic methodology has become more sophisticated, and as new growth-related genes have been discovered, some children with apparent ISS have specific defects in one of many genes along the GH-IGF-1 axis (see Chapter 3).

In 2003 the US FDA approved the administration of GH for children with ISS and height SDS ≤2.25. Although the USA is the only country that has approved the use of GH in ISS, there had been prior longstanding use of this treatment in an off-label fashion in children who had unexplained short stature in many countries, including the USA.

A number of meta-analyses have reviewed the efficacy of rhGH treatment in children with ISS, such studies suggest average height gains between 3 and 7 cm [91, 92]. As with all rhGH treatment indications, responses vary among patients (table 7) [92–97]. In GH trials with children with ISS reported to date, small sample sizes remain a major problem, as well as the relatively high percentage of dropouts of non-responders in studies with longer duration of treatment.

Data from large databases and long-term GH postmarketing studies indicate that rates of adverse events are slightly lower, but not statistically significantly different, in this patient group than in other rhGH-treated patients [98, 99].

SHOX Gene Haploinsufficiency (SHOX)

The SHOX gene is located in the pseudoautosomal regions at the distal ends of the X and Y chromosomes, this gene encodes a homeodomain transcription

Table 7. GH trials in children with ISS

Patients n	Mean age at start, years	Mean height SD at start	GH dose mg/kg/day	Mean duration of treatment years	Estimated height gain SD	Ref.
68	12.5	–2.7	0.031*	4.4	0.5	93
50	10.1	–3.2	0.035–0.053	6.5	0.8	94
80	10.1	–2.7	0.043	5.7	0.7	95
29	7.8	–2.1	0.039–0.078	8.0	1.0	96
126	11.5	–2.7	0.033–0.067	5.9	1.3	97

* GH dose was 0.074 mg/kg three times per week.

Table 8. GH trials in children with SHOX gene haploinsufficiency

Patients n	Mean age at start, years	Mean height SD at start	GH dose mg/kg/day	Mean duration of treatment years	Estimated height gain SD	Ref.
27	7.3	–3.3	0.050	2.0	0.9	105

factor responsible for a significant proportion of long bone growth [100]. SHOX haploinsufficiency is the primary cause of short stature in TS and Leri-Weill dyschondrosteosis and in 2% to 3% of patients with ISS [101]. The prevalence of this condition is estimated to be approximately 1 in 2500 individuals [101]. Without GH treatment the estimated average adult height of patients with SHOX deficiency is estimated to be –2.3 SDS for females and –1.8 SDS for males [102].

Soon after the discovery of the SHOX gene, experience with rhGH started in patients with Leri-Weill dyschondrosteosis and those with non-syndromic SHOX deficiency, but published studies are mostly case reports or small non-controlled trials [103, 104]. The only randomized controlled trial is summarized in table 8 [105].

Despite the little experience with GH treatment of patients with SHOX gene haploinsufficiency, it appears to have a safety profile comparable to that reported in other pediatric indications for which GH has been previously approved.

Noonan Syndrome (NS)

NS is a relatively common multiple congenital anomaly syndrome characterized by typical facial features, short stature and congenital heart defects [106]. The

Table 9. GH trials in children with NS

Patients n	Mean age at start, years	Mean height SD at start	GH dose mg/kg/day	Mean duration of treatment years	Estimated height gain SD	Ref.
18	8.2	−2.9	0.033–0.066	7.5	1.7	111
24	7	−3.2	0.025–0.11	7.6	0.6	112
29	11	−2.7	0.050	6.4	1.3	113

incidence is estimated to be approximately 1/1,000 to 1/2,500 live births, males and females are affected equally and their karyotypes are normal. The inheritance of NS is autosomal dominant with variability in expression [107]. The molecular biology of NS is discussed in Chapter 3.

Short stature is one of the cardinal features of NS; length at birth is within the normal reference range, but during childhood, affected children grow at a slow rate and the median adult height is reported to be below −2 SDS (162.5 cm in men and 152.7 cm in women) [108]. Specific growth charts have been developed for individuals with NS [109]. The underlying cause of short stature is unknown, but pathology in the GH/IGF-1 axis has been reported and similar to patients with TS, decreased hGH sensitivity has been hypothesized [110].

Over the last two decades, few rhGH treatment trials in children with NS have been reported and fewer data presented of near-adult or adult height. The most important trials in which adult height or near-adult height data are reported are summarized in table 9 [111–113].

Studies have demonstrated that rates of adverse events are not different in this patient group than in other GH-treated patients. Although special attention should be paid to cardiac function and regular echocardiograms are recommended, recent studies have shown no changes in cardiac dimensions [114].

Conclusions

Data from the administration of GH to treat children with sort stature resulting from GHD or GH insufficiency have now accrued over 40 years of clinical experience. The use of rhGH for the treatment of GHD remains the primary indication for GH treatment in childhood, but 7 more indications have been approved over the last years.

In all the conditions discussed GH has been shown to increase height velocity leading to progressive normalization of height SDS during childhood and in GHD, TS and ISS, there is some evidence of increase in adult height.

Even though in these conditions rhGH therapy is considered effective in terms of growth and with a good safety profile, additional considerations are needed to responsibly assess the long-term value of the added height increment and to balance expected benefit with economical considerations. The approvals of rhGH therapy for short non-GHD children has validated the notion of GH sensitivity, which gives the opportunity to some children with significant short stature but with normal GH test results to benefit from rhGH and perhaps attain an adult height within the normal range.

References

1 Baumann G: Heterogeneity of growth hormone; in Bercu BB (ed): Basic and Clinical Aspects of Growth Hormone. New York, Plenum Press, 1988, pp 13–31.

2 Bennett LL: Failure of hypophyseal growth hormone to produce nitrogen storage in a girl with hypophyseal dwarfism. J Clin Endocrinol 1950; 10:492–495.

3 Knobil E, Morse A, Wolf RC, Greep RO: The action of bovine, porcine and simian growth hormone preparations on the costochondral junction in the hypophysectomized rhesus monkey. Endocrinology 1958;62:348–354.

4 Levine LS, Sonenberg M, New MI: Metabolic effects in children of a 37 amino acid fragment of bovine growth hormone. J Clin Endocrinol Metab 1973;37:607–615.

5 Li CH, Evans HM: The isolation of pituitary growth hormone. Science 1944;99:183–184.

6 Raben MS: Therapy of a pituitary dwarf with human growth hormone. J Clin Endocrinol Metab 1958;18:901–903.

7 Raben MS: Growth hormone. 2. Clinical use of human growth hormone. N Engl J Med 1962;266: 82–86.

8 Fradkin JE: Creutzfeldt-Jakob disease in pituitary growth hormone recipients. Endocrinologist 1993;3:108–114.

9 Frasier SD: Human pituitary growth hormone therapy in growth hormone deficiency. Endocr Rev 1983;4:155–170.

10 Cohen P, Bright GM, Rogol AD, Kappelgaard A-M, Rosenfeld RG, on behalf of the American Norditropin Clinical Trials Group: Effects of dose and gender on the growth and growth factor response to growth hormone (GH) in GH-deficient children: implications for efficacy and safety. J Clin Endocrinol Metab 2002;87:90–98.

11 Richmond EJ, Rogol AD: Growth hormone deficiency in children. Pituitary 2008;11:115–120.

12 Reiter EO, Price DA, Wilton P, Albertsson-Wikland K, Ranke MB: Effect of growth hormone (GH) treatment on the near-final height of 1,258 patients with idiopathic GH deficiency: analysis of a large international database. J Clin Endocrinol Metab 2006;91:2047–2054.

13 Cutfield WS, Lindberg A, Albertsson-Wikland K, Chatelain P, Ranke MB, Wilton P: Final height in idiopathic growth hormone deficiency: the KIGS experience. KIGS International Board. Acta Paediatr Suppl 1999;88:72–75.

14 Maghnie M, Ambrosini L, Cappa M, Pozzobon G, Ghizzoni L, Ubertini MG, di Iorgi N, Tinelli C, Pilia S, Chiumello G, Lorini R, Loche S: Adult height in patients with permanent growth hormone deficiency with and without multiple pituitary hormone deficiencies. J Clin Endocrinol Metab 2006; 91:2900–2905.

15 Blethen SL, Baptista J, Kuntze J, Foley T, LaFranchi S, Johanson A. Adult height in growth hormone (GH)-deficient children treated with biosynthetic GH. The Genentech Growth Study Group. J Clin Endocrinol Metab 1997;82:418–420.

16 MacGillivray MH, Blethen SL, Buchlis JG, Clopper RR, Sandberg DE, Conboy TA: Current dosing of growth hormone in children with growth hormone deficiency: how physiologic? Pediatrics 1998;102:527–530.

17 August GP, Julius JR, Blethen SL: Adult height in children with growth hormone deficiency who are treated with biosynthetic growth hormone: the National Cooperative Growth Study experience. Pediatrics 1998;102:512–516.

18 Birnbacher R, Riedl S, Frisch H: Long-term treatment in children with hypopituitarism: pubertal development and final height. Horm Res 1998;49: 80–85.

19   Joss E, Zuppinger K, Schwarz HP, Roten H: Final height of patients with pituitary growth failure and changes in growth variables after long-term hormonal therapy. Pediatr Res 1983;17:676–679.

20   De Luca F, Maghnie M, Arrigo T, Lombardo F, Messina MF, Bernasconi S: Final height outcome of growth hormone-deficient patients treated since less than five years of age. Acta Paediatr 1996;85:1167–1171.

21   Wilson TA, Rose SR, Cohen P, Rogol AD, Backeljauw P, Brown R, Hardin DS, Kemp SF, Lawson M, Radovick S, Rosenthal SM, Silverman L, Speiser P, The Lawson Wilkins Pediatric Endocrinology Society Drug and Therapeutics Committee: Update of guidelines for the use of growth hormone in children: the Lawson Wilkins Pediatric Endocrinology Society Drug and Therapeutics Committee. J Pediatr 2003;143:415–421.

22   GH Research Society Consensus Guidelines for the Diagnosis and Treatment of Growth Hormone (GH) Deficiency in Childhood and Adolescence: Summary Statement of the GH Research Society. J Clin Endocrinol Metab 2000;85:3990–3993.

23   Mauras N, Attie KM, Reiter EO, Saenger P, Baptista J and the Genentech Inc Cooperative Study Group: High dose recombinant human growth hormone (GH) treatment of GH-deficient patients in puberty increases near-final height: a randomized, multicenter trial. J Clin Endocrinol Metab 2000;85:3653–3660.

24   Cohen P, Rogol AD, Howard CP, Bright GM, Kappelgaard AM, Rosenfeld RG, American Norditropin Study Group: Insulin growth factor-based dosing of growth hormone therapy in children: a randomized, controlled study. J Clin Endocrinol Metab 2007;92:2480–2486.

25   Ho KK, 2007 GH Deficiency Consensus Workshop Participants: Consensus guidelines for the diagnosis and treatment of adults with GH deficiency II: a statement of the GH Research Society in association with the European Society for Pediatric Endocrinology, Lawson Wilkins Society, European Society of Endocrinology, Japan Endocrine Society, and Endocrine Society of Australia. Eur J Endocrinol 2007;157:695–700.

26   Norman LJ, Macdonald IA, Watson AR: Optimising nutrition in chronic renal insufficiency – growth. Pediatr Nephrol 2004;19:1245–1252.

27   Tönshoff B, Mehls O: Interaction between glucocorticoids and the somatotrophic axis. Acta Paediatr Suppl 1996;417:72–75.

28   Schaefer F, Seidel C, Binding A, Gasser T, Largo RH, Prader A, Schärer K: Pubertal growth in chronic renal failure. Pediatr Res 1990;28:5–10.

29   Vimalachandra D, Hodson EM, Willis NS, Craig JC, Cowell C, Knight JF: Growth hormone for children with chronic kidney disease. Cochrane Database Syst Rev 2006;3:CD003264.

30   Mehls O, Wühl E, Tönshoff B, Schaefer F, Nissel R, Haffner D: Growth hormone treatment in short children with chronic kidney disease. Acta Paediatr 2008;97:1159–1164.

31   Haffner D, Schaefer F, Nissel R, Wühl E, Tönshoff B, Mehls O: Effect of growth hormone treatment on the adult height of children with chronic renal failure. German Study Group for Growth Hormone Treatment in Chronic Renal Failure. N Engl J Med 2000;343:923–930.

32   Crompton CH; Australian and New Zealand Paediatric Nephrology Association: Long-term recombinant human growth hormone use in Australian children with renal disease. Nephrology (Carlton) 2004;9:325–330.

33   Hokken-Koelega A, Mulder P, De Jong R, Lilien M, Donckerwolcke R, Groothof J: Long-term effects of growth hormone treatment on growth and puberty in patients with chronic renal insufficiency. Pediatr Nephrol 2000;14:701–706.

34   Powell DR, Liu F, Baker BK, Hintz RL, Lee PD, Durham SK, Brewer ED, Frane JW, Watkins SL, Hogg RJ: Modulation of growth factors by growth hormone in children with chronic renal failure. The Southwest Pediatric Nephrology Study Group. Kidney Int 1997;51:1970–1979.

35   Fine RN, Kohaut EC, Brown D, Perlman AJ: Growth after recombinant human growth hormone treatment in children with chronic renal failure: report of a multicenter randomized double-blind placebo-controlled study. Genentech Cooperative Study Group. J Pediatr 1994;124:374–382.

36   Bérard E, André JL, Guest G, Berthier F, Afanetti M, Cochat P, Broyer M, on behalf of the French Society for Pediatric Nephrology: Long-term results of rhGH treatment in children with renal failure: experience of the French Society of Pediatric Nephrology. Pediatr Nephrol 2008;23:2031–2038.

37   Fine RN, Stablein D, Cohen AH, Tejani A, Kohaut E: Recombinant human growth hormone post-renal transplantation in children: a randomized controlled study of the NAPRTCS. Kidney Int 2002;62:688–696.

38   Fine RN, Ho M, Tejani A, Blethen S: Adverse events with rhGH treatment of patients with chronic renal insufficiency and end-stage renal disease. J Pediatr 2003;142:539–545.

39 Saenger P, Attie KM, DiMartino-Nardi J, Fine RN: Carbohydrate metabolism in children receiving growth hormone for 5 years. Chronic renal insufficiency compared with growth hormone deficiency, Turner syndrome, and idiopathic short stature. Genentech Collaborative Group. Pediatr Nephrol 1996;10:261–263.

40 Guest G, Bérard E, Crosnier H, Chevallier T, Rappaport R, Broyer M: Effects of growth hormone in short children after renal transplantation. French Society of Pediatric Nephrology. Pediatr Nephrol 1998;12:437–446.

41 Nielsen J, Wohlert M: Chromosome abnormalities found among 34,910 newborn children: results from a 13-year incidence study in Arhus, Denmark. Hum Genet. 1991;87:81–83.

42 Turner HH: A syndrome of infantilism, congenital webbed neck, and cubitus valgus: Endocrinology 1938;23:566–574.

43 Ullrich O: Über typische Kombinationsbilder multipler Abartungen. Z Kinderheilk 1930;49:271–276.

44 Ferguson-Smith MA: Karyotype-phenotype correlations in gonadal dysgenesis and their bearing on the pathogenesis of malformations. J Med Genet 1965;2:142–155.

45 Rochiccioli P, David M, Malpuech G, Colle M, Limal JM, Battin J, Mariani R, Sultan C, Nivelon JL, Simonin G: Study of final height in Turner's syndrome: ethnic and genetic influences. Acta Paediatr 1994;83:305–308.

46 Holl RW, Kunze D, Etzrodt H, Teller W, Heinze E: Turner syndrome: final height, glucose tolerance, bone density and psychosocial status in 25 adult patients. Eur J Pediatr 1994;153:11–16.

47 Rao E, Weiss B, Fukami M, Rump A, Niesler B, Mertz A, Muroya K, Binder G, Kirsch S, Winkelmann M, Nordsiek G, Heinrich U, Breuning MH, Ranke MB, Rosenthal A, Ogata T, Rappold GA: Pseudoautosomal deletions encompassing a novel homeobox gene cause growth failure in idiopathic short stature and Turner syndrome. Nat Genet 1997;16:54–63.

48 Cave CB, Bryant J, Milne R: Recombinant growth hormone for children and adolescents with Turner syndrome. Cochrane Database Syst Rev 2007;1:CD003887.

49 Rosenfeld RG, Attie KM, Frane J, Brasel JA, Burstein S, Cara JF, Chernausek S, Gotlin RW, Kuntze J, Lippe BM, Mahoney CP, Moore WV, Saenger P, Johanson AJ: Growth hormone therapy of Turner's syndrome: beneficial effect on adult height. J Pediatr 1998;132:319–324.

50 Chernausek SD, Attie KM, Cara JF, Rosenfeld RG, Frane J: Growth hormone therapy of Turner syndrome: the impact of age of estrogen replacement on final height. Genentech Inc Collaborative Study Group. J Clin Endocrinol Metab 2000;85:2439–2445.

51 Quigley CA, Crowe BJ, Anglin DG, Chipman JJ, and the US Turner Syndrome Study Group: Growth hormone and low dose estrogen in Turner syndrome: results of a United States multi-center trial to near-final height. J Clin Endocrinol Metab 2002;87:2033–2041.

52 Stephure DK: Canadian Growth Hormone Advisory Committee: Impact of growth hormone supplementation on adult height in Turner syndrome: results of the Canadian randomized controlled trial. J Clin Endocrinol Metab 2005;90:3360–3366.

53 Carel JC, Mathivon L, Gendrel C, Ducret JP, Chaussain JL: Near normalization of final height with adapted doses of growth hormone in Turner's syndrome. J Clin Endocrinol Metab 1998;83:1462–1466.

54 Van Pareren YK, de Muinck Keizer-Schrama SM, Stijnen T, Sas TC, Jansen M, Otten BJ, Hoorweg-Nijman JJ, Vulsma T, Stokvis-Brantsma WH, Rouwé CW, Reeser HM, Gerver WJ, Gosen JJ, Rongen-Westerlaken C, Drop SL: Final height in girls with Turner syndrome after long-term growth hormone treatment in three dosages and low dose estrogens. J Clin Endocrinol Metab 2003;88:1119–1125.

55 Ranke MB, Partsch CJ, Lindberg A, Dorr HG, Bettendorf M, Hauffa BP, Schwarz HP, Mehls O, Sander S, Stahnke N, Steinkamp H, Said E, Sippell W: Adult height after GH therapy in 188 Ullrich-Turner syndrome patients: results of the German IGLU Follow-Up Study 2001. Eur J Endocrinol 2002;147:625–633.

56 Soriano-Guillen L, Coste J, Ecosse E, Léger J, Tauber M, Cabrol S, Nicolino M, Brauner R, Chaussain JL, Carel JC: Adult height and pubertal growth in Turner syndrome after treatment with recombinant growth hormone. J Clin Endocrinol Metab 2005;90:5197–5204.

57 Davenport ML, Crowe BJ, Travers SH, Rubin K, Ross JL, Fechner PY, Gunther DF, Liu C, Geffner ME, Thrailkill K, Huseman C, Zagar AJ, Quigley CA: Growth hormone treatment of early growth failure in toddlers with Turner syndrome: a randomized, controlled, multicenter trial. J Clin Endocrinol Metab 2007;92:3406–3416.

58   Pasquino AM, Pucarelli I, Segni M, Tarani L, Calcaterra V, Larizza D: Adult height in sixty girls with Turner syndrome treated with growth hormone matched with an untreated group. J Endocrinol Invest 2005;28:350–356.

59   Bolar K, Hoffman AR, Maneatis T, Lippe B: Long-term safety of recombinant human growth hormone in Turner syndrome. J Clin Endocrinol Metab 2008;93:344–351.

60   Radetti G, Pasquino B, Gottardi E, Boscolo Contandin I, Aimaretti G, Rigon F: Insulin sensitivity in Turner's syndrome: influence of GH treatment. Eur J Endocrinol 2004;151:351–354.

61   Sas TC, de Muinck Keizer-Schrama SM, Stijnen T, Aanstoot HJ, Drop SL: Carbohydrate metabolism during long-term growth hormone (GH) treatment and after discontinuation of GH treatment in girls with Turner syndrome participating in a randomized dose-response study. Dutch Advisory Group on Growth Hormone. J Clin Endocrinol Metab 2000;85:769–775.

62   Clayton PE, Cianfarani S, Czernichow P, Johannsson G, Rapaport R, Rogol AD: Management of the child born small for gestational age through adulthood: a consensus statement. J Clin Endocrinol Metab 2007;92:804–810.

63   Lee PA, Chernausek SD, Hokken-Koelega AC, Czernichow P: International Small for Gestational Age Advisory Board consensus development conference statement: management of the short child born small for gestational age. Pediatrics 2001;111:1253–1261.

64   Ester W, Bannink E, van Dijk M, Willemsen R, van der Kaay D, de Ridder M, Hokken-Koelega A: Subclassification of small for gestational age children with persistent short stature: growth patterns and response to GH treatment. Horm Res 2008;69:89–98.

65   Gibson AT, Carney S, Cavazzoni E, Wales JK: Neonatal and postnatal growth. Horm Res 2000;53(suppl 1):42–49.

66   Leger J, Noel M, Czernichow P: Growth factors and intrauterine growth retardation. II. Serum growth hormone, insulin-like growth factor (IGF)-I, and IGF-binding protein 3 levels in children with intrauterine growth retardation compared with normal control subjects: prospective study from birth to two years of age. Study Group of IUGR. Pediatr Res 1996;40:101–107.

67   Abuzzahab MJ, Schneider A, Goddard A, Grigorescu F, Lautier C, Keller E, Kiess W, Klammt J, Kratzsch J, Osgood D, Pfäffle R, Raile K, Seidel B, Smith RJ, Chernausek SD, Intrauterine Growth Retardation Study Group: IGF-I receptor mutations resulting in intrauterine and postnatal growth retardation. N Engl J Med 2003;349:2211–2222.

68   Lee PA, Blizzard RM, Cheek DB, Holt AB: Growth and body composition in intrauterine growth retardation before and during human growth hormone administration. Metabolism 1974;23:913–919.

69   Chernausek SD: Treatment of short child born small for gestational age: US perspective. Horm Res 2005;64(suppl 2):63–66.

70   Cooke RW, Foulder-Hughes L: Growth impairment in the very preterm and cognitive and motor performance at 7 years. Arch Dis Child 2003;88:482–487.

71   De Zegher F, Ong KK, Ibañez L, Dunger DB: Growth hormone therapy in short children born small for gestational age. Horm Res 2006;65(suppl 3):145–152.

72   De Zegher F, Hokken-Koelega A: Growth hormone therapy for children born small for gestational age: height gain is less dose dependent over the long term than over the short term. Pediatrics 2005;11:e458–e462.

73   Dahlgren J, Wikland KA: Final height in short children born small for gestational age treated with growth hormone. Pediatr Res 2005;57:216–222.

74   Carel JC, Chatelain P, Rochiccioli P, Chaussain JL: Improvement in adult height after growth hormone treatment in adolescents with short stature born small for gestational age: results of a randomized controlled study. J Clin Endocrinol Metab 2003;88:1587–1593.

75   Van Pareren Y, Mulder P, Houdijk M, Jansen M, Reeser M, Hokken-Koelega A: Adult height after long-term, continuous growth hormone (GH) treatment in short children born small for gestational age: results of a randomized, double-blind, dose-response GH trial. J Clin Endocrinol Metab 2003;88:3584–3590.

76   De Zegher F, Ong K, van Helvoirt M, Mohn A, Woods K, Dunger D: High-dose growth hormone (GH) treatment in non-GH-deficient children born small for gestational age induces growth responses related to pretreatment GH secretion and associated with a reversible decrease in insulin sensitivity. J Clin Endocrinol Metab 2002;87:148–1451.

77 Van Pareren Y, Mulder P, Houdijk M, Jansen M, Reeser M, Hokken-Koelega A: Effect of discontinuation of growth hormone treatment on risk factors for cardiovascular disease in adolescents born small for gestational age. J Clin Endocrinol Metab 2003;88:347–353.

78 Cutfield WS, Lindberg A, Rapaport R, Wajnrajch MP, Saenger P: Safety of growth hormone treatment in children born small for gestational age: the US trial and KIGS analysis. Horm Res 2006; 65(suppl 3):153–159.

79 Nicholls RD: Genomic imprinting and uniparental disomy in Angelman and Prader-Willi syndromes: a review. Am J Med Genet 1993;46: 16–25.

80 Prader A, Labhart A, Willi H: Ein Syndrom von Adipoditas, Kleinwuchs, Kryptorchismus and Oligophrenie. Schweiz Med Wochenschr 1956;86: 1260–1261.

81 Brambilla P, Bosio L, Manzoni P, Pietrobelli A, Beccaria L, Chiumello G: Peculiar body composition in patients with Prader-Labhart-Willi syndrome. Am J Clin Nutr 1997;65:1369–1374.

82 Cassidy SB, Dykens E, Williams CA: Prader-Willi and Angelman syndromes: sister-imprinted disorders. Am J Med Genet 2000;97:136–146.

83 Myers SE, Carrel AL, Whitman BY, Allen DB: Sustained benefit after two years of growth hormone on body composition, fat utilization, physical strength and agility, and growth in Prader-Willi syndrome. J Pediatr 2000;137:42–49.

84 Lindgren AC, Hagenäs L, Müller J, Blichfeldt S, Rosenborg M, Brismar T, Ritzén EM: Growth hormone treatment of children with Prader-Willi syndrome affects linear growth and body composition favourably. Acta Paediatr 1998;87:28–31.

85 Lindgren AC, Lindberg A Growth hormone treatment completely normalizes adult height and improves body composition in Prader-Willi syndrome: experience from KIGS (Pfizer International Growth Database). Horm Res 2008;70: 182–187.

86 Festen DA, de Lind van Wijngaarden R, van Eekelen M, Otten BJ, Wit JM, Duivenvoorden HJ, Hokken-Koelega AC: Randomized controlled growth hormone trial: effects on anthropometry, body composition, and body proportions in a large group of children with Prader-Willi syndrome. Clin Endocrinol (Oxf) 2008;69:443–451.

87 Eiholzer U: Deaths in children with Prader-Willi syndrome. A contribution to the debate about the safety of growth hormone treatment in children with PWS. Horm Res 2005;63:33–39.

88 Allen DB, Carrel AL: Growth hormone therapy for Prader-Willi syndrome: a critical appraisal. J Pediatr Endocrinol Metab 2004;17(suppl 4):1297– 1306.

89 Lindgren AC, Hagenäs L, Ritzén EM: Growth hormone treatment of children with Prader-Willi syndrome: effects on glucose and insulin homeostasis. Swedish National Growth Hormone Advisory Group. Horm Res 1999;51:157–161.

90 Wit JM, Boersma B, de Muinck Keizer-Schrama SM, Nienhuis HE, Oostdijk W, Otten BJ, Delemarre-Van de Waal HA, Reeser M, Waelkens JJ, Rikken B: Long-term results of growth hormone therapy in children with short stature, subnormal growth rate and normal growth hormone response to secretagogues. Dutch Growth Hormone Working Group. Clin Endocrinol (Oxf) 1995;42:365–372.

91 Finkelstein BS, Imperiale TF, Speroff T, Marrero U, Radcliffe DJ, Cuttler L: Effect of growth hormone therapy on height in children with idiopathic short stature: a meta-analysis. Arch Pediatr Adolesc Med 2002;156:230–240.

92 Bryant J, Baxter L, Cave CB, Milne R: Recombinant growth hormone for idiopathic short stature in children and adolescents. Cochrane Database Syst Rev 2007;3:CD004440.

93 Leschek EW, Rose SR, Yanovski JA, Troendle JF, Quigley CA, Chipman JJ, Crowe BJ, Ross JL, Cassorla FG, Blum WF, Cutler GB Jr, Baron J, National Institute of Child Health and Human Development-Eli Lilly & Co. Growth Hormone Collaborative Group: Effect of growth hormone treatment on adult height in peripubertal children with idiopathic short stature: a randomized, double-blind, placebo-controlled trial. J Clin Endocrinol Metab 2004;89:3140–3148.

94 Wit JM, Rekers-Mombarg LT, Cutler GB, Crowe B, Beck TJ, Roberts K, Gill A, Chaussain JL, Frisch H, Yturriaga R, Attanasio AF: Growth hormone treatment to final height in children with idiopathic short stature: evidence for a dose effect. J Pediatr 2005;146:45–53.

95 Hintz RL, Attie KM, Baptista J, Roche A: Effect of growth hormone treatment on adult height of children with idiopathic short stature. Genentech Collaborative Group. N Engl J Med 1999;340:502– 507.

96 Elder CJ, Barton JS, Brook CG, Preece MA, Dattani MT, Hindmarsh PC: A randomised study of the effect of two doses of biosynthetic human growth hormone on final height of children with familial short stature. Horm Res 200812;70:89– 92.

97 Albertsson-Wikland K, Aronson AS, Gustafsson J, Hagenäs L, Ivarsson SA, Jonson B, Kriström B, Marcus C, Nilsson KO, Ritzén EM, Tuvemo T, Westphal O, Åman J: Dose-dependent effect of growth hormone on final height in children with short stature without growth hormone deficiency. J Clin Endocrinol Metab 2008;93:4342–4350.

98 Quigley CA, Gill AM, Crowe BJ, Robling K, Chipman JJ, Rose SR, Ross JL, Cassorla FG, Wolka AM, Wit JM, Rekers-Mombarg LT, Cutler GB: Safety of growth hormone treatment in pediatric patients with idiopathic short stature. J Clin Endocrinol Metab 2005;90:5188–5196.

99 Kemp SF, Kuntze J, Attie KM, Maneatis T, Butler S, Frane J, Lippe B: Efficacy and safety results of long-term growth hormone treatment of idiopathic short stature. J Clin Endocrinol Metab 2005;90:5247–5253.

100 Rao E, Weiss B, Fukami M, Rump A, Niesler B, Mertz A, Muroya K, Binder G, Kirsch S, Winkelmann M, Nordsiek G, Heinrich U, Breuning MH, Ranke MB, Rosenthal A, Ogata T, Rappold GA: Pseudoautosomal deletions encompassing a novel homeobox gene cause growth failure in idiopathic short stature and Turner syndrome. Nat Genet 1997;16:54–63.

101 Jorge AA, Souza SC, Nishi MY, Billerbeck AE, Libório DC, Kim CA, Arnhold IJ, Mendonca BB: SHOX mutations in idiopathic short stature and Leri-Weill dyschondrosteosis: frequency and phenotypic variability. Clin Endocrinol (Oxf) 2007;66:130–135.

102 Ross JL, Kowal K, Quigley CA, Blum WF, Cutler GB Jr, Crowe B, Hovanes K, Elder FF, Zinn AR: The phenotype of short stature homeobox gene (SHOX) deficiency in childhood: contrasting children with Leri-Weill dyschondrosteosis and Turner syndrome. J Pediatr 2005;147:499–507.

103 Munns CF, Berry M, Vickers D, Rappold GA, Hyland VJ, Glass IA, Batch JA: Effect of 24 months of recombinant growth hormone on height and body proportions in SHOX haploinsufficiency. J Pediatr Endocrinol Metab 2003;16:997–1004.

104 Ogata T, Onigata K, Hotsubo T, Matsuo N, Rappold G: Growth hormone and gonadotropin-releasing hormone analog therapy in haploinsufficiency of SHOX. Endocr J 2001;48: 317–322.

105 Blum WF, Crowe BJ, Quigley CA, Jung H, Cao D, Ross JL, Braun L, Rappold G, SHOX Study Group: Growth hormone is effective in treatment of short stature associated with short stature homeobox-containing gene deficiency: two-year results of a randomized, controlled, multicenter trial. J Clin Endocrinol Metab 2007;92:219–228.

106 Noonan JA, Ehmke DA: Associated non-cardiac malformations in children with congenital heart disease. J Pediatr 1963;63:468–470.

107 Nora JJ, Nora AH, Sinha AK, Spangler RD, Lubs HA: The Ullrich-Noonan syndrome (Turner phenotype). Am J Dis Child 1974;127:48–55.

108 Ranke MB, Heidemann P, Knupfer C, Enders H, Schmaltz AA, Bierich JR: Noonan syndrome: growth and clinical manifestations in 144 cases. Eur J Pediatr 1988;148:220–227.

109 Witt DR, Keena BA, Hall JG, Allanson JE: Growth curves for height in Noonan syndrome. Clin Genet 1986;30:150–153.

110 Ahmed ML, Foot AB, Edge JA, Lamkin VA, Savage MO, Dunger DB: Noonan's syndrome: abnormalities of the growth hormone/IGF-I axis and the response to treatment with human biosynthetic growth hormone. Acta Paediatr Scand 1991;80:446–450.

111 Osio D, Dahlgren J, Wikland KA, Westphal O: Improved final height with long-term growth hormone treatment in Noonan syndrome. Acta Paediatr 2005;94:1232–1237.

112 Raaijmakers R, Noordam C, Karagiannis G, Gregory JW, Hertel NT, Sipilä I, Otten BJ: Response to growth hormone treatment and final height in Noonan syndrome in a large cohort of patients in the KIGS database. J Pediatr Endocrinol Metab 2008;21:267–273.

113 Noordam C, Peer PG, Francois I, De Schepper J, van den Burgt, Otten BJ: Long-term growth hormone treatment improves adult height in children with Noonan syndrome with and without mutations in PTPN11. Eur J Endocrinol 2008;159:203–208.

114 Noordam C, Draaisma JM, van den Nieuwenhof J, van der Burgt I, Otten BJ, Daniels O: Effects of growth hormone treatment on left ventricular dimensions in children with Noonan's syndrome. Horm Res 2001;56:110–113.

Alan D. Rogol, MD, PhD
Department of Pediatrics, University of Virginia
685 Explorers Rd, Charlottesville, VA 22911-8441 (USA)
Tel. +1 804 971 6687, Fax +1 804 971 1147
E-Mail adrogol@comcast.net

Hindmarsh PC (ed): Current Indications for Growth Hormone Therapy, ed 2, revised.
Endocr Dev. Basel, Karger, 2010, vol 18, pp 109–125

GH Use in the Transition of Adolescence to Adulthood

Nelly Mauras

Division Endocrinology and Metabolism, Nemours Children's Clinic, Jacksonville, Fla., USA

Abstract

The achievement of final adult height occurs much earlier than the acquisition of peak bone mass and muscle strength in both genders, males achieving these milestones later than females. This *transition* period is particularly challenging in adolescents that were treated with GH during childhood. A high percentage of these adolescents in transition are no longer considered GH-deficient at the completion of their linear growth, making it necessary to retest GH reserves using potent secretagogues (such as insulin tolerance tests (ITT) or arginine/GHRH) and more stringent cutoff peaks (<6 ng/ml in this age group for ITT for example) to establish the correct diagnosis of GH deficiency. This is especially important in those with idiopathic isolated GH deficiency. Who, how and when to test are critical so that patients who are truly GH-deficient can be selected for life-long therapy while others are spared. Data on the efficacy of GH in these patients have shown mixed results (some positive changes, others minimal or no improvement) on body composition and bone mineral density (BMD) accrual, e.g., most likely due to the differences in the length of time that patients have been off GH in the published studies. In those persistently deficient with abnormally low BMD (i.e. <–2 SD), low IGF-1 and abnormal body composition (high % body fat), GH should be continued starting at doses of 12.5 µg/kg·day and IGF-1 measured periodically and kept in the mid to upper limits of normal (within +1 SD). A repeat DEXA scan for BMD and body composition assessment a year after initiation of treatment is useful to guide therapy. GH is no substitute for healthy eating, weight control and regular exercise in diminishing cardiovascular risk, but these subjects should likely be treated at least through their mid-20s when peak BMD and muscle strength are achieved, at which time reassessment of the lifelong need for GH can be made in the idiopathic GH-deficient group. Those with multiple pituitary hormone deficiencies on the other hand should be replaced indefinitely with GH. *The treatment of the GH-deficient adolescent in transition should be individualized.*

Introduction

Many youngsters have been treated with GH for the diagnosis of GH deficiency as children with the predominant goal to achieve a taller height. Until relatively recently, when these patients achieved adult height, it was customary to discontinue GH administration. However, even though the average male finishes growth between 16.5 and 17 years, and females between 14.5 and 15 years, muscle mass and strength, as well as bone mineral density (BMD) do not peak until the mid-20s, males later than females. This period when the child is through growing yet has not achieved the peak muscle and bone development of adulthood has been termed the *transition phase*. For adolescents that were treated with GH during childhood, establishing the correct diagnosis of GH deficiency is critical so that patients who are truly GH-deficient can be selected for lifelong therapy while others are spared. This is an area of great importance in contemporary endocrinology. Over the last few years, a number of studies have shown that the actions of GH are far more complex than stimulating linear growth, with potent effects promoting lipolysis, lean body mass accrual, bone mineralization, normal cardiac function, exercise tolerance, and even quality of life (QOL) all past the years of completion of linear growth [1, 2]. A well-defined profile of GH deficiency syndrome in the adult has evolved, prompting the approval of GH replacement in adults with GH deficiency in many countries.

This chapter examines available data in the literature regarding the need to continue or not GH past the linear growth years in patients treated with GH in childhood. The review will be restricted solely to subjects considered childhood-onset GH-deficient as, so far, there are no data to suggest the need for adult GH replacement in individuals treated with GH for non-GH-deficient reasons, such as patients with Turner syndrome, small for gestational age, chronic renal failure, Prader-Willi syndrome or idiopathic short stature for example.

Who, When and How Do We Retest for GH Deficiency during Transition?

Who to Test?
It has become increasingly evident that a large proportion of children with childhood-onset GH deficiency may not be considered GH-deficient as adults, especially those whose deficiency in GH is isolated and idiopathic. Only 40–75% of adolescents with isolated GH deficiency have persistent GH deficiency on retesting [3–6] suggesting the need to repeat provocative testing in all children with either idiopathic GH deficiency or with documentation of only one other pituitary hormone deficiency.

On the other hand, children with confirmed organic pituitary disease causing multiple pituitary hormone deficiencies (MPHD), or individuals with confirmed mutations in genes controlling somatotroph development or GH gene expression (combined pituitary hormone deficiency), or those with midline or pituitary structural anomalies have a much greater likelihood to have persistent GH deficiency and may not require retesting. It has been suggested that the positive predictive value for GH deficiency in adults with deficiencies of three pituitary hormones is 96%, and 99% in adults with four pituitary deficiencies [7]. This high probability of persistent GH deficiency is also true for adolescents in the transition age with MPHD or a structurally abnormal pituitary on repeat magnetic resonance imaging [8]. Interestingly, in a study of adolescents with a history of cranial irradiation and GH deficiency, 50% did not meet the criteria for adult GH deficiency on retesting [9] indicating that these patients should also be retested. Hartman et al. [7] have suggested that in individuals with MPHD or those with high probability of persistent deficiency, that GH is discontinued for 1 month after the completion of linear growth and that serum IGF-1 concentration is measured. If the serum IGF-1 concentration is <−2 SD from the mean, GH provocative testing is not needed and GH therapy could be continued in lower doses (see below). This approach has been also favored by others [10, 11].

When to Test?
How long patients who have been treated with GH as children, some for many years, need to be off GH before retesting is not fully defined. Guidelines from the GH Research Society recent consensus meeting and clinical practice guidelines [1, 12] have recommended from 1 to 3 months off GH therapy prior to retesting. This strategy is useful in clinical practice and should be considered a minimum interval. In patients with other pituitary hormone deficiencies, adequate replacement in those hormones should be documented prior to retesting.

An important and provocative study by Italian investigators [13] suggested that in those identified as isolated GH-deficient prior to the onset of puberty, that the retesting should not wait until after completion of adult height, but be done in mid-puberty. These authors found that one-third of those retested in mid-puberty their GH peaked to >10 ng/ml, hence GH was discontinued and that their final adult height was not significantly different than those considered persistently deficient who continued GH therapy until adult height. These data, if confirmed by other studies, could effect future recommendations on the timing for retesting. This however awaits further study.

How to Test?
The definition of what constitutes GH deficiency based on provocative testing has remained an area of intense controversy in endocrinology as the cutoffs chosen are often arbitrary, depending on the potency of the secretagogue used and the sensitivity of the GH assay. It is however clear that GH production rates vary greatly depending on age, with peak GH production coinciding with peak height velocity in children [14]. GH production remains relatively robust until the mid-20s then progressively declines with age and relative hyposomatropism can be observed in

the elderly [15]. Because of this physiologic variation with age, the same standards applied to adults in terms of expected GH production should not be applied to adolescents in the transition phase.

Insulin-induced hypoglycemia (insulin tolerance test, ITT) remains the gold standard test for detecting GH deficiency [1, 12, 16] and should be the preferred choice when retesting adolescents in the transition period [10–12]. In adults, the peak GH response is considered normal in an ITT if >3 ng/ml [16], however this cutoff is clearly too restrictive in the adolescents in transition. Maghnie et al. [17] studied the GH responses to ITT in young subjects with known pituitary disease and compared their data with those of healthy controls (ages 15–30). They observed that a cutoff of normal >6.1 ng/ml has 96% sensitivity and 100% specificity to detect GH deficiency in the transition phase. To date these represent the best data on the cutoffs for ITT responses in this age group.

The GHRH-arginine test has been proposed as a reasonable alternative to the ITT [11, 18], with a sensitivity and specificity similar to insulin-induced hypoglycemia. Biller et al. [18] showed that a peak GH concentration of 4.1 ng/ml for GHRH-arginine using a two-site immunochemiluminescent assay achieved a sensitivity of 95% and specificity of 92% for detecting GH deficiency. To achieve a specificity of 95%, the threshold for the ITT was 3.3 ng/ml and 1.5 ng/ml for the GHRH-arginine test. This test however is not fully reliable in those with suspected hypothalamic disease [19].

Serum IGF-1 concentrations have not been too helpful and are not conventionally used as a criterion for GH deficiency unless the concentrations are low. Maghnie et al. [8, 17] suggested that an IGF-1 SDS of <−1.7 is diagnostic of GH deficiency in the transition phase but serum IGF-1 concentrations may not fall below normal levels until 6 months after GH is discontinued [8]. Serum IGF-1 concentrations are however quite helpful in MPHD patients as those with profoundly low IGF-1 as defined above may not need repeat provocative testing.

Efficacy Data

Previous multicenter trials have examined the use of GH in this GH-deficient adolescent population in the transition phase using several study protocol models and yielded somewhat mixed results [3–5, 20–24]. Lean body mass and bone mass accrual were shown to be better in those severe GH-deficient patients in whom GH therapy was continued versus those in whom it was discontinued, but individuals who had been off GH a mean of 1.3 years (range 6 weeks to 5 years) were shown to still accrue bone mass very well after reinitiation of treatment [5]. Table 1 summarizes current study data.

Table 1. Efficacy data in the GH-deficient patients in transition

	Vahl et al. [20]	Carroll et al. [23], Drake et al. [4]	Underwood et al. [22]	Shalet et al. [5], Attanasio [24]	Mauras et al. [6]
Patients, n	19	24	64	149	58
Age, years	16–26	14–20	23.8±4.2	19	15.8±1.8
Dx	13 idiopathic	14 MPHD	12 idiopathic	80% MPHD	18 idiopathic
Prior GH Rx, years	>3	>5	>1	>1	>3
Time off GH	Not stated	On GH at recruitment	>5 years	6 weeks to 5 years	GH at recruitment
Dose	0.006 µg/ m²·day	11.6 µg/kg·day	12.5 or 25 µg/kg·day	12.5 or 25 µg/kg·day	20 µg/kg·day
Study type	PC, DB RCT × 1 year, followed by GH × 1 year in all	RCT, (+)GH or (–)GH × 1 year. Baseline done after 3 months (+)GH or (–)GH	PC, DB, RCT with low- or high-dose GH × 2 years	PC, DB, RCT with low- or high-dose GH × 2 years	PC, DB, RCT × 2 years. Those that failed ITT (n = 18) were controls
Body composition	No increase in LBM in either group, increase in % FM in PL not in GH	FM: no difference between groups LBM: (+)GH: +2.5 kg, none (–)GH	FM: PL +2.3 kg, low GH –0.7 kg, high GH –3.7 kg LBM: low and high GH: +13.4% both, PL +3.7%	FM: low GH –6%, –7% high GH LBM: (+) GH low and high (14.2%/12.7%)	No difference in the 3 groups (GH, PL or controls) in FM or LBM at 24 months
BMD		(+)GH: +6% (–)GH: +2.4% p = 0.074 between groups	Increased in 25 µg > 12.5 > PL	Increased both, no dose effect	No difference in the 3 groups (GH, PL or controls)
Muscle strength	No difference				No difference
LV Function	N/A		No difference in ECHOs		No difference in stress test or ECHOs
Lipids	No difference	No difference	No difference	No difference, but LDL/HDL ratio (+) GH	No difference
Glucose/insulin	No difference	Better insulin sensitivity in controls	No difference		No difference
QOL	No difference		No difference		No difference
Comment	Key outcomes at 12 months	Key outcomes at 12 months	Key outcomes at 24 months	Key outcomes at 24 months	Key outcomes at 24 months

Body Composition

The beneficial effects of GH replacement therapy in adults have been well characterized [1, 2, 25–27], including a decrease in percentage body fat and increase in lean body mass, both of which are associated with a decrease in risk of cardiovascular disease. A small cohort of 11 adolescent subjects with GH deficiency had a decrease in resting energy expenditure and an increase in adiposity within a few weeks after discontinuation of GH, effects sustained for 1 year of observation and not detected in healthy controls [28]. In 19 adolescents with childhood-onset GH deficiency, Vahl et al. [20] studied them at the time of discontinuation of treatment, followed them for 1 year on either placebo or GH and then an additional year on GH in all subjects. They observed a significant increase in percentage body fat after 1 year in the placebo-treated group and no change in the GH-treated group. Attanasio et al. [21], in a large cohort of multinational subjects, showed a significant decrease in adiposity with GH treatment in adolescents in transition but no dose effect (12.5 vs. 25 µg/kg•day), yet a marked gender effect, with much more robust responses in males. Similar results were observed by Underwood, et al. [22] (fig. 1a–d). Johannsson et al. [29] reported no changes in percentage body fat over 2 years in healthy controls, whereas persistently GH-deficient subjects had a more significant increase in percentage fat mass 2 years after discontinuation of treatment than those who discontinued treatment but upon retesting were considered GH-sufficient.

We performed a double-blind randomized placebo-controlled trial in 58 adolescents who were treated with GH in childhood for GH deficiency and who had nearly completed growth at study entry [6]. The primary objective was to establish the efficacy regarding body composition and BMD changes, as well as the safety, of a transition dose (20 µg/kg•day). As secondary objectives, we studied the effects of GH treatment on plasma lipids, IGF-1 concentration, carbohydrate metabolism, cardiac function, exercise tolerance, and QOL. The experimental design also allowed us to further evaluate the frequency of persistent GH deficiency after retesting at the time of therapy cessation. All baseline studies (DEXA, hand-grip dynamometry, hormones and substrates, carotid intima thickness and echocardiography) were performed while the subjects were still on GH therapy. Subsequently all subjects discontinued GH for 6 weeks and an ITT was performed. Patients were considered to have persistent GH deficiency if the peak GH responses were <5 ng/ml, while subjects who were GH-sufficient at retesting served as untreated controls. Those considered still GH-deficient were randomized to GH or placebo injections for 2 years. Study testing was repeated at year 1 and year 2.

In this study, all three groups, GH-treated and placebo-treated GH-deficient subjects as well as controls, had a significant increase in percentage fat mass over the 2 years of the study (fig. 2). Although the increase was less in the GH-treated group than those treated with placebo, these differences were not sustained at 24 months. This lack of improvement after 2 years could be a result of the sustained decrease in

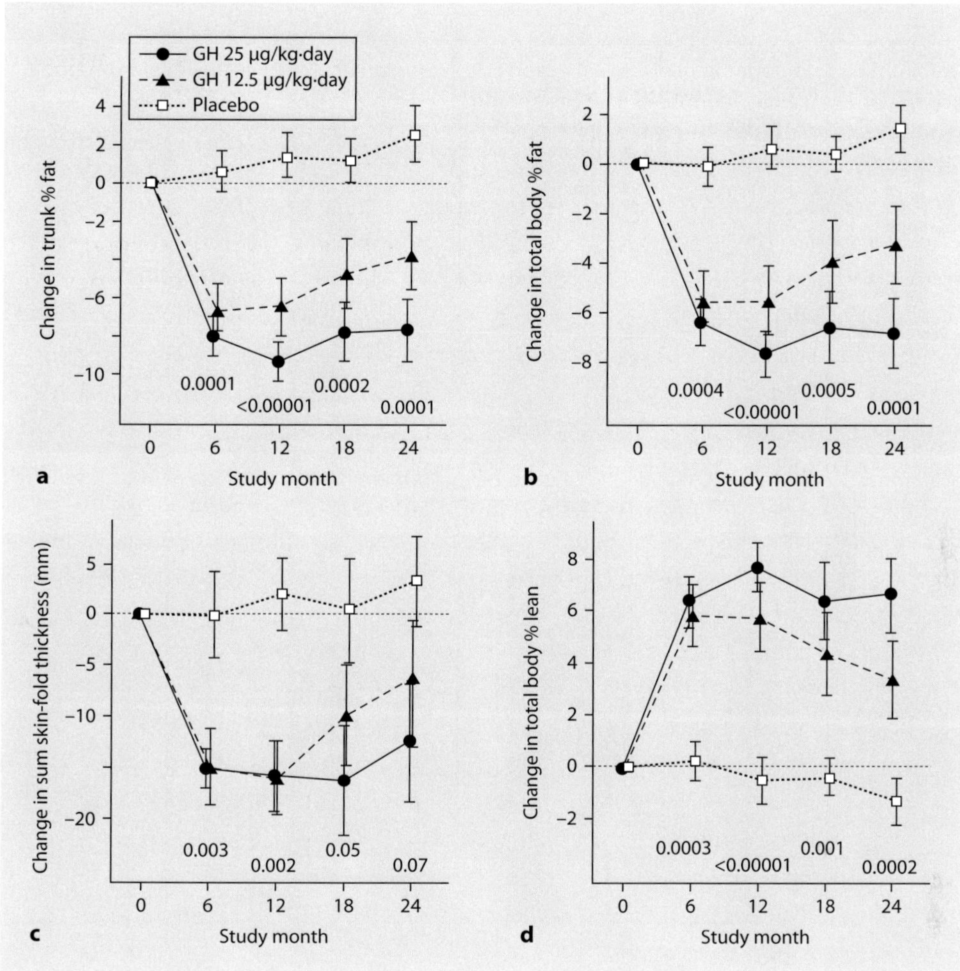

Fig. 1. **a–d** Changes (mean ± SE) from baseline in trunk percentage fat (DEXA), total percentage fat, and total percentage lean mass for patients with baseline and 24-month data treated by Underwood et al. [22] with GH at two different doses or placebo for 2 years [reproduced with permission].

GH dosing (relative to the dose given during active growth). Overall, GH therapy seems to improve lean body mass accrual in those GH-deficient subjects in transition who have been off GH for the longest period, but not in those who achieved normal body composition through puberty, at least for 2 years post discontinuation.

Bone Mineral Density
Previous studies have suggested that adults with GH deficiency, both adult- and childhood-onset, have lower BMD than healthy controls [30–33]. The bone-

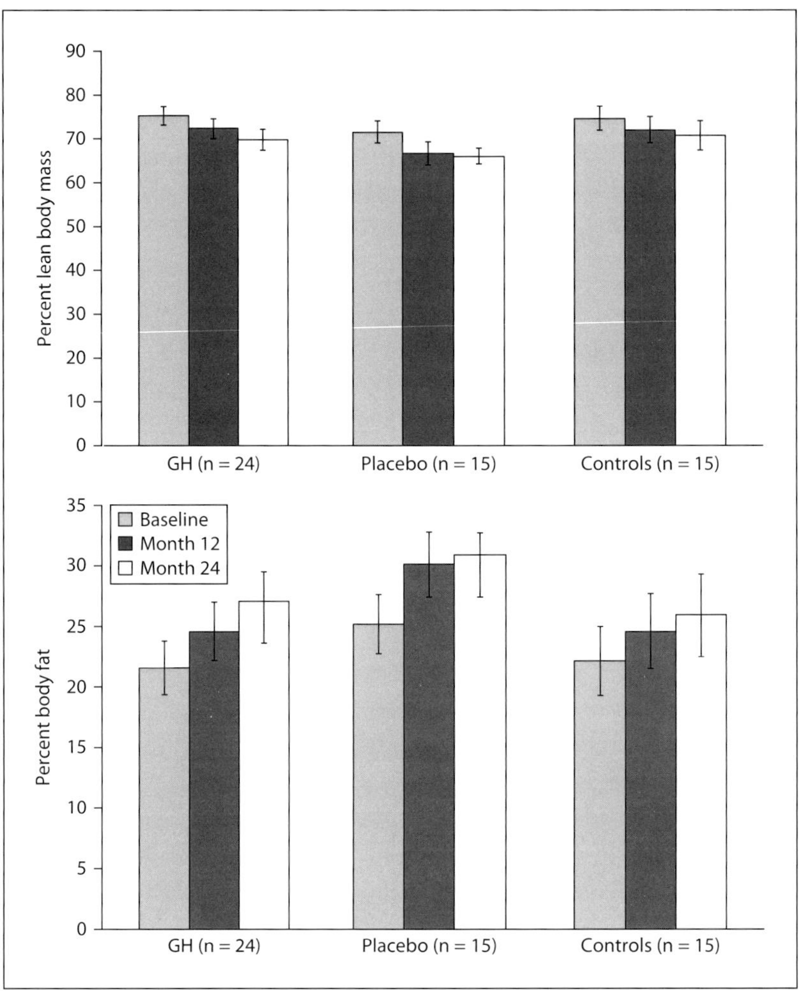

Fig. 2. Changes in lean body mass and percentage fat mass in GH-deficient adolescents in transition [for further details, see 6]. There were no significant differences between the three treatment groups.

anabolic actions of GH have been extensively studied, and the impact of GH on bone mass accrual can continue even after discontinuation of therapy for over 1.5 years [34]. In an open-label trial, Drake et al. [4] randomized GH-deficient adolescents in the transition phase to GH treatment or observation and recorded a greater percentage of BMD accrual with GH. Additionally, continuation of GH therapy for 2 years in adolescents in transition resulted in greater bone mass accrual rates in those treated with higher GH doses (25 µg/kg•day), although those on placebo also accrued bone mass [5]. In another study [6] no differences in BMD accrual response in GH-deficient patients treated with GH versus those that

Mauras

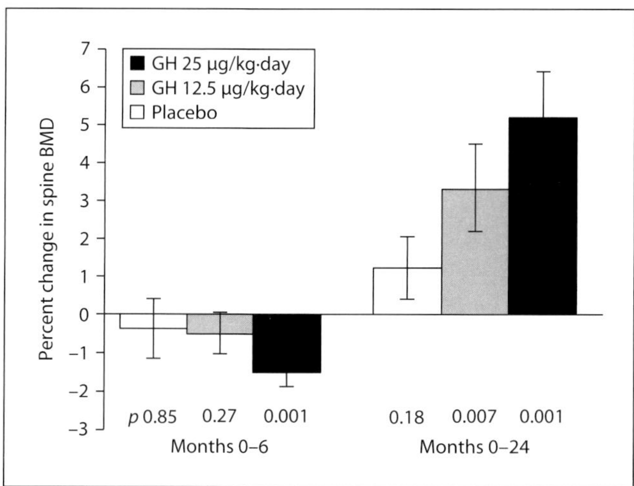

Fig. 3. Positive changes described in those patients off GH the longest. Percent change from baseline in spine BMD for those patients with baseline and month 24 data (mean ± SE). Dose response: months 0–6; p = 0.035; months 0–24, p = 0.018 [22, reproduced with permission].

received placebo after 2 years. However, compared with a healthy, non-GH-treated population, the Z scores at baseline were normal in both the GH-deficient and the control group. These data suggest that GH therapy in GH-deficient children can normalize bone mass in adolescence and that the accrual of bone mass continues after the completion of linear growth, even without GH, for at least 2 years. In aggregate, continuation of GH therapy during transition may be most crucial in those who have MPHD or in those who's GH has been discontinued the longest. Figure 3 exemplifies some of the positive changes described in those patients off GH the longest.

Lipids/Carbohydrate Metabolism

Data thus far have offered inconsistent results regarding the effects of GH on plasma lipid concentrations, both in adults and in adolescents [1, 2]. Several studies have shown no improvement in lipid markers after either discontinuation or reinitiation of GH therapy in this period [20, 21, 29]. These data are different from those reported by Colao et al. [35] in a much smaller cohort of subjects, but congruent with the lack of sustained benefit after 24 months observed in adults with childhood-onset GH deficiency treated with GH [6, 22].

Much of the reported data in this age group has shown no change in any measure of carbohydrate metabolism or in global measures of insulin sensitivity in GH-treated versus placebo-treated and control subjects [6, 7, 20, 22, 23, 29]. GH therapy, however, can be associated with mild hyperinsulinemia yet normal blood glucose concentrations. Using hyperinsulinemic euglycemic clamps in adolescents with GH deficiency in the transition phase, Nørrelund et al. [3] showed a significant increase in insulin sensitivity in the placebo-treated group compared with the GH-treated group, suggesting that continued GH treatment does not overcome the insulin-antagonistic effects of GH over time. Collectively, the data thus far are neutral regarding carbohydrate metabolism and GH should not be withheld because of this concern.

Muscle Strength

GH has not been shown to increase muscle strength in the majority of studies examining this question [12, 36–38]. Administration of GH or withdrawal of GH in the cohort of GH-deficient adolescents in transition reported by us was not accompanied by any significant changes in skeletal muscle strength [6]. Although our measure of strength (hand-grip) may well be inadequate to demonstrate subtle changes in efficacy, it was an easy-to-use, reproducible tool, convenient for a multicenter trial. The results are, however, similar to those reported in adults both short and long term [22, 27] and similar to those observed in trials of younger adolescent subjects [20]. Data thus far do not support the use of GH for the purpose of increasing muscle strength in this age group.

Echocardiography/Exercise

Data in adults has not shown a significant benefit of GH enhancing cardiac function or exercise performance. However, recent data from multicenter trials suggest that although administration of GH to GH-deficient adults for 32 weeks does not change submaximal exercise performance, it improved maximal exercise capacity (as measured by Vo_2max) without any influence of physical activity [37]. Reports from trials using GH in the transition to adulthood have shown no substantial changes in cardiac function in childhood-onset GH-deficient adults, with or without treatment [22]. We also showed no significant detrimental effects of GH therapy on cardiac parameters of resting systolic and diastolic function in the treatment group [6]. These are some of the most detailed data to date in this age group, including carotid intima thickness, cardiac echocardiography and exercise tolerance. However, no significant beneficial effects could be shown over the 24-month period either. The data are reassuring and suggest that GH-deficient children replaced with GH (i.e. using GH doses of 0.3 mg/kg•week) have normal cardiac function at the completion of linear growth. Colao et al. [35] showed a mild decrease in cardiac function in a group of GH-deficient adolescents in transition; however, the group was small and uncontrolled, and the cardiac parameters were still within the normal range.

Quality of Life

The results of studies on QOL in GH deficiency have been conflicting, some showing an improvement of some aspects of QOL, whereas others show no difference with GH replacement [20, 39–57]. When specific tools designed for use in patients with GH deficiency were used, GH replacement appears to improve QOL measures [40]; however, these findings appear to be mostly in adult-onset GH-deficient states. In the placebo-controlled randomized trial we conducted in adolescents in transition [6] we included serial measurements of several dimensions of QOL that have proven sensitive to GH deficiency or GH replacement therapy in other studies, such as general health-related QOL, social adjustment, and aspects of QOL that are specific to GH deficiency [58]. The effects of GH replacement therapy on QOL were minimal in our study [6], similar to other reports [22]. Interestingly, the study sample had basal QOL scores that were indistinguishable from those of the general population, which indirectly suggest that several years of GH treatment in childhood has positive effects on QOL. Taken in aggregate, data thus far do not show an improvement in measures of QOL in GH-deficient adolescents in transition to adulthood as compared to control groups.

Dose

The average dose of rhGH in the USA is ~42 µg/kg•day for replacement during childhood, and often doses as high as 100 µg/kg•day may be used during puberty [59]. The issue of proper dosing is important in the GH-deficient adolescent in transition, and it became clear that if GH was to be continued, the doses of GH used should be intermediate between those used during childhood treatment and adult replacement. This concept is also consistent with the fact that GH production rates markedly decline after the achievement of adult height but are higher than those of older adults [15]. Adolescents and young adults with GH deficiency have received a wide range of GH doses during the transition period, lower than childhood treatment yet higher than adult replacement. Doses ranging from 12.5, 20 or 25 µg/kg•day have been used is several studies in the transition phase. Those receiving doses of 25 µg/kg•day performed better than those receiving 12.5 µg/kg•day in some studies [22], but there was not a clear dose effect in other studies [5, 24]. Clayton et al. [10] have recommended using absolute doses instead, from 0.2 to 0.5 mg/day (~3 to 7 µg/kg•day in a 70-kg individual). I personally prefer the use of 12.5 µg/kg•day as a starting dose based on the data published thus far, titrating as needed to keep the IGF-1 concentrations in the upper limits for age. This is also the approach suggested by Radovick and DiVall [11] in a recent review of the topic.

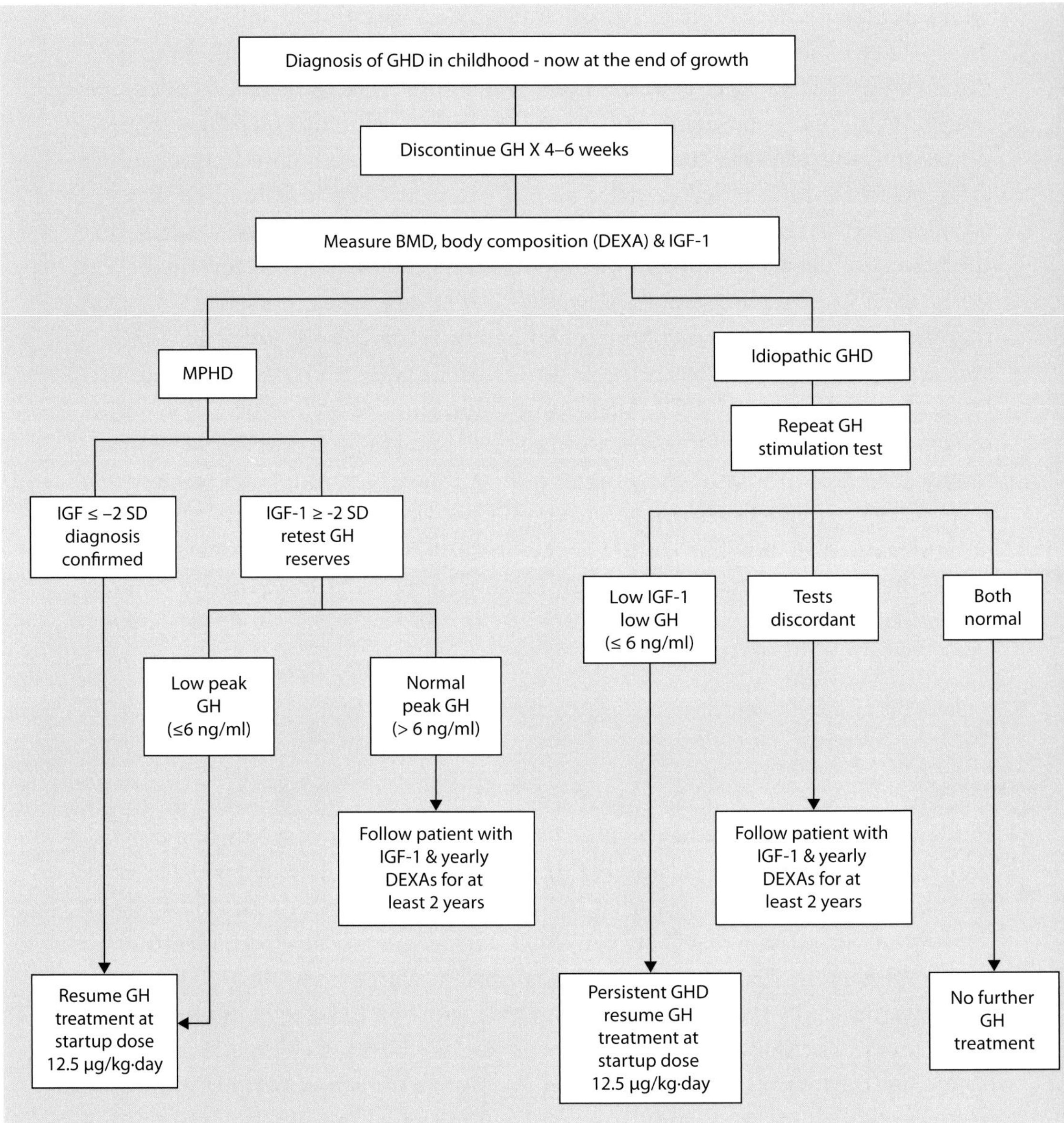

Fig. 4. Flowchart for diagnosis and treatment of adolescents with GHD in transition.

Follow-Up

These patients are often not only in transition metabolically, but emotionally and intellectually. Many go off to college/university in this period and over the following 5 years making major decisions regarding training, employment and relationships. Many also are mobile, transferring to different cities and losing track of the endocrinologists of their youth. It is critically important that care is organized so that replacement therapy is continued in those that stay on GH during the transition phase, and that those that retest GH-deficient yet discontinue treatment are closely followed for several years. Because pediatric endocrinologists have followed these patients often for years, I believe it is the obligation of the pediatric specialist to reassess pituitary function at the end of growth and help the patient make appropriate decisions regarding long-term therapy. The transition to adult care should only happen after those decisions have been made and therapeutic choices properly established. A baseline assessment of BMD and body composition with DEXA and a serum IGF-1 concentration should be done before reinitiation of GH therapy. This testing should also be done in those with MPHD who continue GH uninterrupted but in much lower doses. Fasting lipids can be obtained as well, but as discussed above, the utility of lipid concentrations in assessing need for GH is questionable. Serum IGF-1 concentrations should be measured at every visit and the GH dose titrated to keep concentrations in the mid to upper limits for age (0 to +1 SD). DEXA can be repeated yearly, and depending on the normality of the results, interval between repeat scans can be extended at the clinician's discretion.

Conclusions

Many GH-deficient patients properly treated in childhood can have normal BMD, body composition, cardiac function, muscle strength, and measures of carbohydrate and lipid metabolism as well as measures of QOL by the time they reach adult height. Hence, the evaluation and treatment of the GH-deficient adolescent that completes his/her linear growth requires careful reassessment of GH secretory status during the transition phase to adulthood before we commit them to GH therapy for life. A brief summary of the management is summarized in figure 4. The management is different in those with MPHD vs. those who are only missing GH and in whom no organic pathology was ever documented (idiopathic). Retesting of the GH reserves with potent secretagogues (such as ITT) and well-validated assays is essential especially in those with only one or two hormonal deficiencies. Stringent criteria for diagnosis of persistent GH deficiency should be used (i.e., serum GH concentration <6 μg/ml during an ITT). In those persistently deficient with abnormally low BMD (i.e., <–2 SD), low serum IGF-1 concentration and abnormal body composition (high percentage body fat), GH should be continued starting at doses of 12.5 μg/kg•day and serum IGF-1 concentrations measured periodically and kept in the mid to upper limits of normal (0 to +1 SD). A repeat DEXA scan for BMD and body composition assessment a year after initiation of treatment is useful to guide therapy. GH is no substitute for healthy eating, weight control and regular exercise in diminishing cardiovascular risk, but these subjects should likely be treated at least through their mid-20s when peak BMD and muscle strength are achieved, at which time reassessment of the lifelong need for GH can be made in the idiopathic GH-deficient group.

References

1   Molitch ME, Clemmons DR, Malozowski S, Merriam GR, Shalet SM, Vance ML; Endocrine Society's Clinical Guidelines Subcommittee, Stephens PA: Evaluation and treatment of adult growth hormone deficiency: an Endocrine Society Clinical Practice Guideline. J Clin Endocrinol Metab 2006;91:1621–1634.

2   Vance ML, Mauras N: Growth hormone therapy in adults and children. N Engl J Med 1999;341:1206–1216.

3   Nørrelund H, Vahl N, Juul A, Møller N, Alberti KG, Skakkebæk NE, Christiansen JS, Jørgensen JO: Continuation of growth hormone (GH) therapy in GH-deficient patients during transition from childhood to adulthood: impact on insulin sensitivity and substrate metabolism. J Clin Endocrinol Metab 2000;85:1912–1917.

4   Drake WM, Carroll PV, Maher KT, Metcalfe KA, Camacho-Hübner C, Shaw NJ, Dunger DB, Cheetham TD, Savage MO, Monson JP: The effect of cessation of growth hormone (GH) therapy on bone mineral accretion in GH-deficient adolescents at the completion of linear growth. J Clin Endocrinol Metab 2003;88:1658–1663.

5   Shalet SM, Shavrikova E, Cromer M, Child CJ, Keller E, Zapletalova J, Moshang T, Blum WF, Chipman JJ, Quigley CA, Attanasio AF: Effect of growth hormone (GH) treatment on bone in postpubertal GH-deficient patients: a 2-year randomized, controlled, dose-ranging study. J Clin Endocrinol Metab 2003;88:4124–4129.

6   Mauras N, Pescovitz OH, Allada V, Messig M, Wajnrajch MP, Lippe B, Transition Study Group: Limited efficacy of growth hormone (GH) during transition of GH-deficient patients from adolescence to adulthood: a phase III multicenter, double-blind, randomized two-year trial. J Clin Endocrinol Metab 2005;90:3946–3955.

7   Hartman ML, Crowe BJ, Biller BM, Ho KK, Clemmons DR, Chipman JJ; HyposCCS Advisory Board; US HypoCCS Study Group: Which patients do not require a GH stimulation test for the diagnosis of adult GH deficiency? J Clin Endocrinol Metab 2002;87:477–485.

8   Maghnie M, Strigazzi C, Tinelli C, Autelli M, Cisternino M, Loche S, Severi F: Growth hormone (GH) deficiency (GHD) of childhood onset: reassessment of GH status and evaluation of the predictive criteria for permanent GHD in young adults. J Clin Endocrinol Metab 1999;84:1324–1328.

9   Gleeson HK, Gattamaneni HR, Smethurst L, Brennan BM, Shalet SM: Reassessment of growth hormone status is required at final height in children treated with growth hormone replacement after radiation therapy. J Clin Endocrinol Metab 2004;89:662–666.

10 Clayton PE, Cuneo RC, Juul A, Monson JP, Shalet SM, Tauber M, European Society of Paediatric Endocrinology: Consensus statement on the management of the GH-treated adolescent in transition to adult care. Eur J Endocrinol 2005;152:165–170.

11 Radovick S, DiVall S: Approach to the growth hormone-deficient child during transition to adulthood. J Clin Endocrinol Metab 2007;92:1195–1200.

12 Ho KK; GH Deficiency Consensus Workshop Participants 2007 Consensus guidelines for the diagnosis and treatment of adults with GH deficiency II: A statement of the GH Research Society in association with the European Society for Pediatric Endocrinology, Lawson Wilkins Society, European Society of Endocrinology, Japan Endocrine Society, and Endocrine Society of Australia. Eur J Endocrinol 2007;157:695–700.

13 Zucchini S, Pirazzoli P, Baronio F, Gennari M, Bal MO, Balsamo A, Gualandi S, Cicognani A: Effect on adult height of pubertal growth hormone retesting and withdrawal of therapy in patients with previously diagnosed growth hormone deficiency. J Clin Endocrinol Metab 2006;91:4271–4276.

14 Martha PM, Rogol AD, Veldhuis JD, Kerrigan JR, Goodman DW, Blizzard RM: Alterations in the pulsatile properties of circulating growth hormone concentrations during puberty in boys. J Clin Endocrinol Metab 1989;69:563–570.

15 Veldhuis JD, Roemmich JN, Richmond EJ, Rogol AD, Lovejoy JC, Sheffield-Moore M, Mauras N, Bowers CY: Endocrine control of body composition in infancy, childhood, and puberty. Endocr Rev 2005;26:114–146.

16 Ho KK, Hoffman DM: Defining growth hormone deficiency in adults. Metabolism 1995;44(suppl 4):91–96.

17 Maghnie M, Aimaretti G, Bellone S, Bona G, Bellone J, Baldelli R, de Sanctis C, Gargantini L, Gastaldi R, Ghizzoni L, Secco A, Tinelli C, Ghigo E: Diagnosis of GH deficiency in the transition period: accuracy of insulin tolerance test and insulin-like growth factor-I measurement. Eur J Endocrinol 2005;152:589–596.

18 Biller BM, Samuels MH, Zagar A, Cook DM, Arafah BM, Bonert V, Stavrou S, Kleinberg DL, Chipman JJ, Hartman ML: Sensitivity and specificity of six tests for the diagnosis of adult GH deficiency. J Clin Endocrinol Metab 2002;87:2067–2079.

19 Darzy KH, Aimaretti G, Wieringa G, Gattamaneni HR, Ghigo E, Shalet SM: The usefulness of the combined growth hormone (GH)-releasing hormone and arginine stimulation test in the diagnosis of radiation-induced GH deficiency is dependent on the post-irradiation time interval. J Clin Endocrinol Metab 2003;88:95–102.

20 Vahl N, Juul A, Jørgensen JO, Ørskov H, Skakkebæk NE, Christiansen JS: Continuation of growth hormone (GH) replacement in GH-deficient patients during transition from childhood to adulthood: a two-year placebo-controlled study. J Clin Endocrinol Metab 2000;85:1874–1881.

21 Attanasio AF, Howell S, Bates PC, Frewer P, Chipman J, Blum WF, Shalet SM: Body composition, IGF-1 and IGFBP-3 concentrations as outcome measures in severely GH-deficient (GHD) patients after childhood GH treatment: a comparison with adult onset GHD patients. J Clin Endocrinol Metab 2002;87:3368–3372.

22 Underwood LE, Attie KM, Baptista J, Genentech Collaborative Study Group: Growth hormone (GH) dose-response in young adults with childhood onset GH deficiency: a two-year, multicenter, multiple-dose, placebo-controlled study. J Clin Endocrinol Metab 2003;88:5273–5280.

23 Carroll PV, Drake WM, Maher KT, Metcalfe K, Shaw NJ, Dunger DB, Cheetham TD, Camacho-Hübner C, Savage MO, Monson JP: Comparison of continuation or cessation of growth hormone (GH) therapy on body composition and metabolic status in adolescents with severe GH deficiency at completion of linear growth. J Clin Endocrinol Metab 2004;89:3890–3895.

24 Attanasio AF, Shavrikova E, Blum WF, Cromer M, Child CJ, Paskova M, Lebl J, Chipman JJ, Shalet SM, Hypopituitary Developmental Outcome Study Group: Continued growth hormone (GH) treatment after final height is necessary to complete somatic development in childhood-onset GH-deficient patients. J Clin Endocrinol Metab 2004;89:4857–4862.

25 Bengtsson BÅ, Abs R, Bennmarker H, Monson JP, Feldt-Rasmussen U, Hernberg-Stahl E, Westberg B, Wilton P, Wuster D: The effects of treatment and the individual responsiveness to growth hormone (GH) replacement therapy in 665 GH-deficient adults. KIMS Study Group and the KIMS International Board. J Clin Endocrinol Metab 1999;84:3929–3935.

26 Christiansen JS, Jorgensen JO: Beneficial effects of GH replacement therapy in adults. Acta Endocrinol 1991;125:7–13.

27 Mauras N, O'Brien KO, Welch S, Rini A, Helgeson K, Vieira NE, Yergey AL: Insulin-like growth factor 1 and growth hormone (GH) treatment in GH-deficient humans: differential effects on protein, glucose, lipid, and calcium metabolism. J Clin Endocrinol Metab 2000;85:1686–1694.

28 Cowan FJ, Evans WD, Gregory JW: Metabolic effects of discontinuing growth hormone treatment. Arch Dis Child 1999;80:517–523.

29 Johannsson G, Albertsson-Wikland K, Bengtsson BÅ: Discontinuation of growth hormone (GH) treatment: metabolic effects in GH-deficient and GH-sufficient adolescent patients compared with control subjects. Swedish Study Group for Growth Hormone Treatment in Children. J Clin Endocrinol Metab 1999;84:4516–4524.

30 Kaufmann JM, Taelman P, Vermeulen A, Vandeweghe M: Bone mineral status in growth hormone-deficient males with isolated and multiple pituitary deficiencies of childhood onset. J Clin Endocrinol Metab 1992;74:118–123.

31 Holmes SJ, Economou G, Whitehouse RW, Adams JE, Shalet SM: Reduced bone mineral density in patients with adult onset growth hormone deficiency. J Clin Endocrinol Metab 1994;78:669–674.

32 Degerblad M, Bengstsson BA, Bramnert M, Johnell O, Manhem P, Rosen T, Thoren M: Reduced bone mineral density in adults with growth hormone (GH) deficiency: increased bone turnover during 12 months of GH substitution therapy. Eur J Endocrinol 1995;133:180–188.

33 Janssen YJ, Hamdy NA, Frolich M, Roelfsema F: Skeletal effects of two years of treatment with low physiological doses of recombinant human growth hormone (GH) in patients with adult-onset GH deficiency. J Clin Endocrinol Metab 1998;83:2143–2148.

34 Biller BM, Sesmilo G, Baum HB, Hayden D, Schoenfeld D, Klibanski A: Withdrawal of long-term physiological growth hormone (GH) administration: differential effects on bone density and body composition in men with adult-onset GH deficiency. J Clin Endocrinol Metab 2000;85:970–976.

35 Colao A, Di Somma C, Salerno M, Spinelli L, Orio F, Lombardi G: The cardiovascular risk of GH-deficient adolescents. J Clin Endocrinol Metab 2002;87:3650–3655.

36 Yarasheski KE, Zachwieja JJ, Campbell JA, Bier DM: Effect of growth hormone and resistance exercise on muscle growth and strength in older men. Am J Physiol 1995;268:E268–E276.

37 Hartman ML, Weltman A, Zagar A, Qualy RL, Hoffman AR, Merriam GR: Growth hormone replacement therapy in adults with growth hormone deficiency improves maximal oxygen consumption independently of dosing regimen or physical activity. J Clin Endocrinol Metab 2008;93:125–130.

38 Gotherstrom G, Bengtsson BA, Sunnerhagen KS, Johsnnsson G, Svensson J: The effects of five-year growth hormone replacement therapy on muscle strength in elderly hypopituitary patients. Clin Endocrinol 2005;62:105–113.

39 Hull KL, Harvey S: Growth hormone therapy and quality of life: possibilities, pitfalls and mechanisms. J Endocrinol 2003;179:311–313.

40 Wirén L, Johannsson G, Bengtsson BA: A prospective investigation of quality of life and psychological well-being after the discontinuation of GH treatment in adolescent patients who had GH deficiency during childhood. J Clin Endocrinol Metab 2001;86:3494–3498.

41 Björk S, Jönsson B, Westphal O, Levin JE: Quality of life of adults with growth hormone deficiency: a controlled study. Acta Paediatr Scand Suppl 1989;356:55–59.

42 Rosén T, Wirén L, Wilhelmsen L, Wiklund I, Bengtsson BÅ: Decreased psychological well-being in adult patients with growth hormone deficiency. Clin Endocrinol 1994;40:111–116.

43 Badia X, Lucas A, Sanmarti A, Roset M, Ulied A: One-year follow-up of quality of life in adults with untreated growth hormone deficiency. Clin Endocrinol 1998;49:765–771.

44 Abs R, Bengtsson BÅ, Hernberg-Stahl E, Monson JP, Tauber JP, Wilton P, Wuster C: GH replacement in 1,034 growth hormone-deficient hypopituitary adults: demographic and clinical characteristics, dosing and safety. Clin Endocrinol 1999;50:703–713.

45 Sanmarti A, Lucas A, Hawkins F, Webb SM, Ulied A: Observational study in adult hypopituitary patients with untreated growth hormone deficiency (ODA Study). Socio-economic impact and health status. Collaborative ODA (Observational GH Deficiency in Adults) Group. Eur J Endocrinol 1999;141:481–489.

46 McGauley GA: Quality of life assessment before and after growth hormone treatment in adults with growth hormone deficiency. Acta Paediatr Scand Suppl 1989;356:70–72.

47 Bengtsson BÅ, Edén S, Lönn L, Kvist H, Stokland A, Lindstedt G, Bosaeus I, Tolli J, Sjostrom L, Isaksson OG: Treatment of adults with growth hormone (GH) deficiency with recombinant human GH. J Clin Endocrinol Metab 1993;76:309–317.

48 Mårdh G, Lundin K, Borg G, Jonsson B, Lindberg A: Growth hormone replacement therapy in adult hypopituitary patients with growth hormone deficiency: combined data from 12 European placebo-controlled clinical trials. Endocrinol Metab 1994;1(suppl A):43–49.

49 Burman P, Broman JE, Hetta J, Wiklund I, Erfurth EM, Hagg E, Karlsson FA: Quality of life in adults with growth hormone (GH) deficiency: response to treatment with recombinant human GH in a placebo-controlled 21-month trial. J Clin Endocrinol Metab 1995;80:3585–3590.

50 Attanasio AF, Lamberts SW, Matranga AM, Birkett MA, Bates PC, Valk NK, Hilsted J, Bengtsson BA, Strasburger CJ: Adult growth hormone (GH)-deficient patients demonstrate heterogeneity between childhood onset and adult onset before and during human GH treatment. Adult Growth Hormone Deficiency Study Group. J Clin Endocrinol Metab 1997;82:82–88.

Mauras

51 Cuneo RC, Judd S, Wallace JD, Perry-Keene D, Burger H, Lim-Tio S, Strauss B, Stockigt J, Topliss D, Alford F, Hew L, Bode H, Conway A, Handelsman D, Dunn S, Boyages S, Cheung NW, Hurley D: The Australian Multicenter Trial of Growth Hormone (GH) Treatment in GH-Deficient Adults. J Clin Endocrinol Metab 1998;83:107–116.

52 Wirén L, Bengtsson BÅ, Johannsson G: Beneficial effects of long-term GH replacement therapy on quality of life in adults with GH deficiency. Clin Endocrinol 1998;48:613–620.

53 Murray RD, Skillicorn CJ, Howell SJ, Lissett CA, Rahim A, Smethurst LE, Shalet SM: Influences on quality of life in GH deficient adults and their effect on response to treatment. Clin Endocrinol 1999;51:565–573.

54 Wallymahmed ME, Foy P, Shaw D, Hutcheon R, Edwards RH, MacFarlane IA: Quality of life, body composition and muscle strength in adult growth hormone deficiency: the influence of growth hormone replacement therapy for up to 3 years. Clin Endocrinol 1997;47:439–446.

55 Baum HB, Katznelson L, Sherman JC, Biller BM, Hayden DL, Schoenfeld DA, Cannistraro KE, Klibanski A: Effects of physiological growth hormone (GH) therapy on cognition and quality of life in patients with adult-onset GH deficiency. J Clin Endocrinol Metab 1998;83:3184–3189.

56 Florkowski CM, Stevens I, Joyce P, Espiner EA, Donald RA: Growth hormone replacement does not improve psychological well-being in adult hypopituitarism: a randomized crossover trial. Psychoneuroendocrinology 1998;23:57–63.

57 Sandberg DE, MacGillivray MH, Clopper RR, Fung C, LeRoux L, Alliger DE: Quality of life among formerly treated childhood-onset growth hormone-deficient adults: a comparison with unaffected siblings. J Clin Endocrinol Metab 1998;83:1134–1142.

58 Stabler B: Impact of growth hormone (GH) therapy on quality of life along the lifespan of GH-treated patients. Horm Res 2001;56(suppl 1): 55–58.

59 Mauras N, Attie KM, Reiter EO, Saenger P, Baptista J: High-dose recombinant human growth hormone (GH) treatment of GH-deficient patients in puberty increases near-final height: a randomized, multicenter trial. Genentech Inc Cooperative Study Group. J Clin Endocrinol Metab 2000; 85:3653–3660.

Nelly Mauras, MD
Division Endocrinology and Metabolism, Nemours Children's Clinic
807 Children's Way, Jacksonville, FL 32207 (USA)
Tel. +1 904 390 3674, Fax +1 904 858 3948
E-Mail nmauras@nemours.org

Author Index

Subject Index